W9-AJM-099

FLOOR
FRAMING

CHARLEY G. CHADWICK
THOMAS S. COLVIN
GEORGE W. SMITH, JR.

**American Association for
Vocational Instructional Materials**
The National Institute for Instructional Materials
120 Driftmier Engineering Center
Athens, Georgia 30602

The American Association for Vocational Instructional Materials (AAVIM)
is a non-profit national institute.

The institute is a cooperative effort of universities, colleges and divisions
of vocational and technical education in the United States to provide for
excellence in instructional materials.

Direction is given by a representative from each of the states, provinces
and territories. AAVIM also works closely with teacher organizations,
government agencies and industry.

AAVIM Staff

Richard M. Hylton	*Executive Director*
George W. Smith, Jr.	*Production Coordinator*
James E. Wren	*Art Director*
Karen Seabaugh	*Secretary*
Laura Ebbert	*Business Office*
Rhonda Grimes	*Order Department*
Robin Ambrose	*Photocomposition*
Marilyn MacMillan	*Art Staff*
Dean Roberts	*Shipping Department*

Floor Framing Editorial Staff

Lois G. Harrington	*Editor*
George W. Smith, Jr.	*Associate Editor/Art Director*
Allyn Jenkins	*Typography*
Robin Ambrose	
Marilyn MacMillan	*Production Art*

The information in this publication has been compiled from authoritative
sources. Every effort has been made to attain accuracy; however, AAVIM
and its associated individuals are not responsible for misapplication or
misinterpretation of this information and cannot assume liability for the
safety of persons using this publication.

An Equal Opportunity/Affirmative Action Institution

Printed in the United States of America

ISBN 0–89606–261–9

Authors

Charley G. Chadwick, Thomas S. Colvin and George W. Smith,
Jr. are credited with writing of this publication. Mr. Chadwick
is an experienced builder and developer and is the Building
Trades Teacher at Calhoun High School, Calhoun, Georgia.

Mr. Colvin is former Research and Development Specialist of
AAVIM. He has since retired.

Mr. Smith is a former Illustrator and Art Director of AAVIM. He
currently serves as Production Coordinator, AAVIM.

Reviewers

Tim Milner and Paul Tilton served as reviewers of this publica-
tion. Mr. Milner is a builder and serves as Building Trades
Teacher at Hart County Comprehensive High School, Hartwell,
Georgia. Mr. Tilton is a builder and contractor who resides near
Colbert, Georgia.

Contents

Introduction 5

Objectives 6

A. Lumber and Materials Used for Floor Framing 7
1. Lumber Seasoning Methods 7
2. Softwood Grading Systems 7
3. Measuring and Figuring Lumber 9
4. Plywood 11
5. Other Wood Panel Products 12
6. Chemically Treated Lumber 13
7. Nails and Other Fasteners Used in Framing Floors 14

B. Types of Floor Framing 17
1. Platform Framing 17
2. Two-Story Framing 18

C. Selecting and Estimating Flooring Materials 19
1. Selecting Materials for Floor Framing 19
2. Estimating Lumber for Floor Framing 21
3. Selecting Materials for Subfloors 22
4. Estimating Lumber for Subfloors 23
5. Estimating Plywood or Other Panel Products for Subfloors 23

D. Installing Termite Shields and Sill Plates 25
1. Installing Termite Shields 25
2. Installing Sill Plates 26

E. Installing the Perimeter Sill 31

F. Constructing and Installing Built-Up Wood Girders 33
1. Building and Installing a Dropped Girder 33
2. Building and Installing a Butt-to-Girder 35

G. Installing Floor Joists 39
1. Laying Out the Floor Joists 39
2. Cutting and Installing the Floor Joists 42

H. Installing Floor Bridging 45
1. Solid Wood Bridging 45
2. Cross or Herringbone Cross Bridging 46
3. Metal Cross Bridging 47

I. Floor Framing over a Basement 49
1. Providing Outside Support for Joists and
   Girders over a Basement 50
2. Providing Support for the Center Ends of
   Joists over a Basement 50

J. Framing Floor Openings 53
1. Laying Out the Floor Opening 53
2. Installing Framing for Floor Opening 54

K. Installing Subfloors 57
1. Installing Plywood Subflooring 57
2. Installing Other Panel Products 59

Appendix 61

Index 63

Acknowledgements 64

List of Tables

I. Classifications and Grades of
   Common Structural Lumber 8
II. Nominal and Actual Dimensions of Commonly
    Used Pieces of Lumber (Dry) 10
III. Veneer Grades 12
IV. Allowable Spans — Floor Joists 20

FIGURE 1. A house plan and a building permit are two items you need to have before construction begins.

The purpose of this publication is to teach the carpenter/student the skills involved in constructing the floor of a conventional platform framed house.

To begin study of floor frame construction, there are several things that must be assumed. First, it is assumed that a house plan has been selected, a building site has been determined, and preliminary procedures, such as clearing the site and obtaining permits, have been accomplished (Figure 1). It is also assumed that a foundation of concrete and/or masonry block supported by a concrete footing has been completed.

A properly constructed foundation is the first step in building a house. Besides being square, this foundation must be strong enough to support the weight of the house. A foundation usually consists of footings, walls, and piers (Figure 2).

Two types of flooring systems are common in residential construction. One is a wood frame floor; the other is a concrete slab-at-grade floor known as a "slab floor."

If your building has a slab floor, there will be no floor framing work involved on a single-story house unless you build the forms for the slab. Wood floor framing for slab houses is limited to the upper floors. The building of a wood frame floor is the subject of this publication.

Upon completion of this study unit, you will be able to frame floors for common types of residential platform construction.

Framing floors is discussed under the following headings:

A. Lumber and Materials Used for Floor Framing
B. Types of Floor Framing
C. Selecting and Estimating Flooring Materials

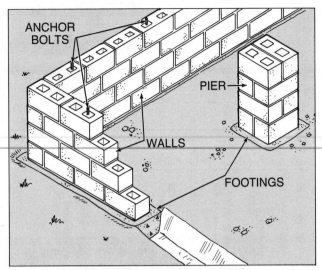

FIGURE 2. The foundation of a conventional platform framed house consists of a concrete footing and concrete block or formed walls and piers.

D. Installing Termite Shields and Sill Plates
E. Installing the Perimeter Sill
F. Constructing and Installing Built-Up Wood Girders
C. Installing Floor Joists
H. Installing Floor Bridging
I. Floor Framing over a Basement
J. Framing Floor Openings
K. Installing Subfloors

OBJECTIVES

Upon successful completion of this study unit, you will be able to do the following:

- Define terms associated with lumber and panel products
- Describe how lumber is seasoned and graded
- Measure and figure lumber
- Define terms associated with floor framing
- Select appropriate materials for floor framing
- Estimate flooring materials
- Install termite shield
- Install sill plates
- Install perimeter sill
- Construct a built-up wood girder
- Install a built-up wood girder
- Install floor joists
- Install floor briding
- Provide needed support over a basement
- Construct framing for floor openings
- Install subflooring

A. Lumber and Materials Used for Floor Framing

The primary material used in floor framing construction is lumber. To meet an increasing demand for lumber, many products have been developed from residue—tree chips, bark, trimming, sawdust, and shavings. Among these products are nonveneered panels and reconstituted sheets, which in **structural** grades, are acceptable materials for subflooring and decking.

While space does not allow a long discussion of lumber and wood panel products, some understanding of these materials is essential to the student.

Lumber and materials used in floor framing are discussed under the following headings:

1. Lumber Seasoning Methods
2. Softwood Grading Systems
3. Measuring and Figuring Lumber
4. Plywood
5. Other Wood Panel Products
6. Chemically Treated Lumber
7. Nails and Other Fasteners Used in Framing Floors

1. LUMBER SEASONING METHODS

There are two methods of seasoning lumber—**air drying** and **kiln drying.**

Air-dried lumber has been stacked in the open air, with wood strips placed between the layers so that air can circulate around each individual board. Several months can be required to dry wood sufficiently to be used in floor framing. Drying will take place faster in hot, dry weather. Look for a grade stamp of "S-Dry," indicating a moisture content of 19 percent or less.

Kiln-dried lumber has been seasoned in a large, temperature-controlled oven called a kiln. It is more expensive than air-dried lumber and will be grade stamped with "KD" or "kiln-dried." Lumber that is grade stamped "KD-15" indicates a moisture content of 15 percent.

After the lumber is seasoned, it can be finished in a planing mill by machines that smooth off the sides and edges. The planed lumber is then examined and graded.

2. SOFTWOOD GRADING SYSTEMS

Lumber grades are based on the quality of the wood. Factors including strength, appearance, moisture content, and the particular species of wood affect grading. Generally speaking, the higher the grade, the fewer knots and other defects. Lower grades will have splits, separations, bark, checks, possible warpage, and a number of loose knots. Grade stamps can furnish valuable information about the recommended use and performance to be expected from a particular piece of lumber.

The American Lumber Standards Committee is responsible for the supervision of the regional lumber inspection organizations that do the actual lumber grading. Sample grade stamps and brief explanations of the marks and symbols found on western and southern woods are shown in Figure 3.

One of the more important bits of information on the grade marking has to do with the moisture content. The drier the wood, the more stable it will be. Wood with a moisture content of 19 percent or above is more likely to warp, twist, or shrink as it dries out.

Classification and grading assigned to lumber is based on its anticipated use, as shown in Table I. Lumber intended for studs, rafters, and joists is **stress-graded**— graded according to the strength of the lumber, not its appearance. The actual qualities of the wood (warps, knots, wavy grain, etc.) determine the strength and the stress grading of lumber.

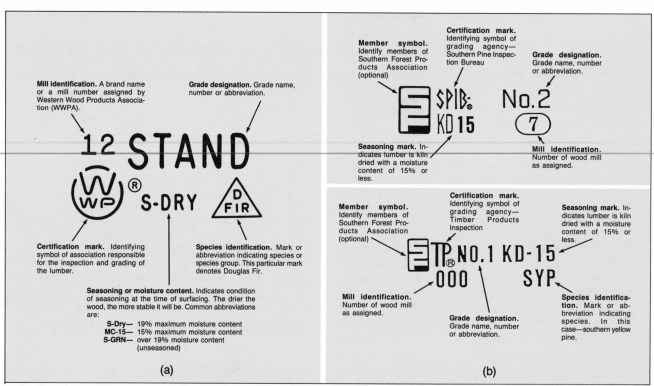

FIGURE 3. Typical grade stamps and information found on (a) western wood and (b) southern pine lumber (two different marks).

(a) Western Wood Products Association, *Product Use Manual 6,* Page 3, August 1987; (b) Southern Forest Products Association, 1988.

Table I
Classification and Grades of Common Structural Lumber
Lengths of lumber generally 6′ and longer, in multiples of 2

Use	Thickness (nominal)	Width (nominal)	Grades
Light Framing	2″ to 4″	2″ to 4″	Construction, Standard, Utility
Studs	2″ to 4″	2″ to 6″ (10′ and shorter)	Stud
Structural Light Framing	2″ to 4″	2″ to 4″	Select structural #1, #2, and #3
Structural Joists & Planks	2″ to 4″	5″ and wider	Select structural #1, #2, and #3
Posts & Timbers	5″ x 5″ (and larger)	**Not** more than 2″ greater than thickness	Select structural #1, #2, and #3
Beams & Stringers	5″ and thicker	More than 2″ greater than thickness	Select structural #1, #2, and #3

Western Woods Products Association, *Product Use Manual,* Page 4, August 1987.

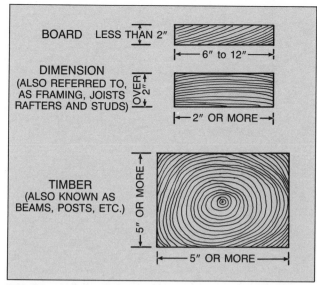

FIGURE 4. Structural softwood lumber is classified in three ways.

For more information about lumber and grading, contact The *National Forest Products Association,* 1250 Connecticut Avenue, N.W., Washington, D.C. 20036; *Southern Forest Products Association,* P.O. Box 52468, New Orleans, Louisiana 70152; and *Western Wood Products Association,* Yeon Building, 522 S.W. Fifth Avenue, Portland, Oregon 97204.

Before you purchase lumber, check your specifications and local building codes to identify the requirements for the lumber you will use in your intended projects.

Structural softwood lumber is size classified in three ways:[1]

Board — defined as being less than 2″ in nominal thickness and 6″ or more wide. If it is under 6″ wide, it may be classified as a strip (Figure 4).

Dimension — from 2″ to (but not including) 5″ thick and is 2″ or more in width. Dimension lumber is also classified as framing, joists, planks, rafters, studs, and small timbers (Figure 4).

Timbers — defined as being 5″ or more in the least dimension. Timbers are also referred to as beams, stringers, posts, caps, girders, purlins, and sills (Figure 4).

---

1. This information adapted from *Grading Rules,* Section 113, Size Classifications, Southern Pine Inspection Bureau (Pensacola, Florida, 1977)

3. MEASURING AND FIGURING LUMBER

When ordering lumber from the building supply company, you need to know the quantity of lumber needed and the size and length of the pieces.

Lumber dimensions are generally referred to by **nominal** dimensions, which differ from the **actual** dimensions (Figure 5). A 2″ x 4″ does not actually measure 2″ thick by 4″ wide.

The reason for this difference lies with the changes that take place after the rough-sawn 2″ x 4″ piece is cut from the tree at the sawmill. First, there are the size-reducing drying processes previously discussed. The dimensions are further reduced when the dried lumber is surfaced at the planing mill. (See lumber surfacing abbreviations in appendix.)

By the time the lumber arrives at the point of sale, the **actual** size of a 2″ x 4″ is 1½″ x 3½″. Because the piece was 2″ x 4″ when cut, it is still referred to as a 2″ x 4″. Table II gives information on the nominal and actual sizes of commonly used boards, planks, and timbers.

NOTE: Pricing of lumber is always figured based on nominal measurements.

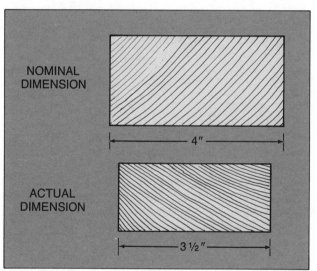

FIGURE 5. The nominal and actual dimensions of lumber are different.

Table II
Nominal and Actual Dimensions of Commonly Used Pieces of Lumber (Dry)

Nominal	Actual
Boards	
1″ x 4″	¾″ x 3½″
1″ x 6″	¾″ x 5½″
1″ x 8″	¾″ x 7½″
1″ x 10″	¾″ x 9¼″
1″ x 12″	¾″ x 11¼″
Dimension Lumber	
2″ x 4″	1½″ x 3½″
2″ x 6″	1½″ x 5½″
2″ x 8″	1½″ x 7¼″
2″ x 10″	1½″ x 9¼″
2″ x 12″	1½″ x 11¼″
4″ x 4″	3½″ x 3½″
4″ x 6″	3½″ x 5½″
4″ x 8″	3½″ x 7½″
4″ x 10″	3½″ x 9¼″
4″ x 12″	3½″ x 11¼″
Timbers	
6″ x 6″	5½″ x 5½″
6″ x 8″	5½″ x 7½″
6″ x 10″	5½″ x 9¼″
6″ x 12″	5½″ x 11¼″

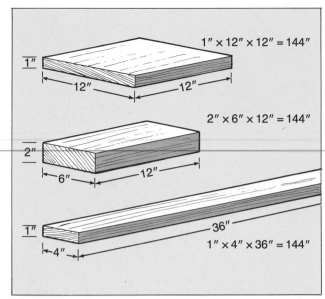

FIGURE 7. A board foot is a unit of quantity for lumber equal to any piece of lumber with a volume of 144 cubic inches.

Generally, softwood lumber is sold in even foot lengths ranging from 6′ to 20′. Longer pieces are available on order at an increased cost.

Lumber is sold by the **board foot.** A board foot is a unit of quantity for lumber equal to any piece of lumber 1″ x 12″ x 12″ or any other measurement with a volume of **144 cubic inches** (Figure 7).

To find out how many board feet there are in a particular piece of lumber, use the following formula. Remember "**T**" = thickness (in inches), "**W**" = width (in inches), and "**L**" = length (in feet). You do not need to convert feet to inches before multiplying.

Formula	Example
$\dfrac{T \times W \times L}{12}$ = board feet	$\dfrac{2″ \times 10″ \times 12′}{12}$ = 20 board feet

For example, the 2″ x 10″ x 12′ piece of lumber mentioned earlier would contain 20 board feet (B.F.).

When constructing a floor, many pieces of lumber of the same size will be ordered. If, for example, the floor you are constructing will utilize eighty-six 2″ x 8″ x 12′ pieces of lumber, use the following formula:

$$\frac{\text{Number of pieces} \times T \times W \times L}{12} =$$

$$\frac{86 \times 2 \times 8 \times 12}{12} = 1,376 \text{ board feet}$$

To determine the cost for this lumber, multiply the number of board feet (1,376) by the cost per board foot (assume a cost of $500 per thousand B.F., or 50¢ per foot).

```
    1,376
  x  .50
    0000
    6880
  $688.00 Total cost for 1,376 board feet
```

Measurements of lumber are always given in the following order: thickness, width, and length (Figure 6). Remember that although the actual thickness and width you get will be smaller, you will request the nominal measurement. **The length will always be the actual length.** For example, a piece of lumber 2″ thick (nominal) by 10″ wide (nominal) by 12′ long (actual) will actually measure 1½″ x 9½″ x 12′ (or just over), but it will be called a 2″ x 10″ x 12′.

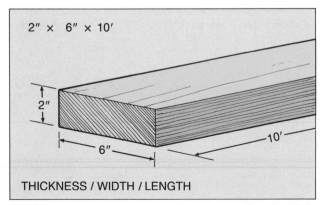

FIGURE 6. Lumber measurement is always given in thickness, width, and length.

Before you compute your lumber cost, you should check current prices with your supplier. Prices for lumber will vary from one area to another. Other variables include grade, length, species, and the supplier.

Some mills and building supply stores price lumber by the piece rather than the board foot. If this is the case, determine the number of pieces needed of a particular size and length. As in the previous example, we will assume you are building a floor requiring eighty-six 2″ x 8″ x 12′ pieces. Assume the price per piece is $8.00. To determine the total cost, use this formula:

Number of pieces needed x cost per piece = total cost

or

86 x $8.00 = $688.00

To find the cost per board foot of an individual piece, first use the formula T x W x L ÷ 12 to determine the number of board feet. For example:

$$\frac{2 \times 8 \times 12}{12} = 16 \text{ B.F.}$$

Take the number of board feet in this piece (16 B.F.) and divide into the cost per piece ($8.00).

$$\frac{\$8.00}{16} = 50¢ \text{ per B.F.}$$

or $500 per thousand B.F.

---

4. PLYWOOD

Plywood is one of the building materials most widely used for structural purposes. It has almost replaced board lumber for subflooring of wood-framed floors.

There are a number of advantages to using plywood rather than board lumber. It has less tendency to warp than boards (provided plywood is stored and applied properly), it can be nailed closer to the edge without splitting, and it can be installed more quickly.

Plywood is made of thin layers of wood called **veneers** that have been peeled off specially selected logs. The log is literally unrolled like a roll of paper towels.

In some plywood grades (sanded underlayment), small defects, such as knotholes and splits, are removed by die cutters, and patches are inserted. After patching, the sheets are glued and sandwiched together (Figure 8). Construction grade panels, commonly known as CDX, are made of 3 and 4 plies and may have small knotholes and limited splits.

It should be noted that this panel is frequently mistaken as an exterior panel. While exterior glue is used in the manufacturing of this panel, it is an interior-type plywood—not to be exposed to water for many days. It should not be used in applications for which it is not designed—namely those that require considerable resistance to weather.

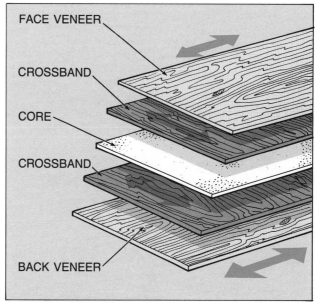

FACE VENEER

CROSSBAND

CORE

CROSSBAND

BACK VENEER

FIGURE 8. Plywood is made up of thin layers of woods glued together with the grains at right angles. It is stronger than boards of the same thickness.

These panels are always made up of an odd number of layers. Each layer consists of one or more plies (single veneer sheets). They are placed so that each layer's grain runs at right angles to the adjacent layer. This procedure is called **cross-laminating.** The outside veneers are referred to as **face** and **back veneers,** the center materials are the **crossbands** and the **core.**

The cross-lamination of layers is what gives plywood a strength that surpasses that of boards of the same thickness. These sandwiches of layers and glue are placed on racks in a hot press. The sheets of veneer are bonded together by a combination of heat and pressure. After panels are laminated and glued, they are trimmed to 4′ x 8′ or 4′ x 10′ sheets and stamped with a grade marking. Sheets measuring 4′ wide x 8′ long are considered to be standard size.

Panel Grading. The factors that affect grading include the appearance of the face and back panels, the extent of knots and splits, and the size and number of patches (repairs). Names stamped on the panel also suggest the panel's intended use (rated "sheathing," "rated Sturd-I-floor," etc.).

The fewer the detracting factors, the higher the grade. If only one side is to be exposed, as in the case of subflooring, a veneer of a lower grade may be used on the unexposed side, and the grade stamp will generally appear on that side.

Typical trademarks of performance-rated panels and a brief explanation of terms and applications can be seen in Figure 9. Table III provides a description of common veneer grades.

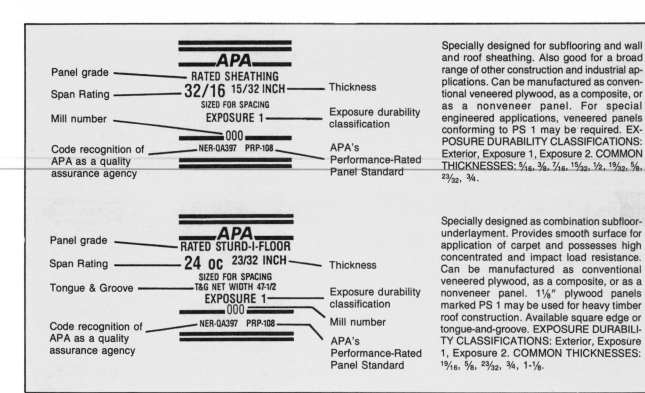

Specially designed for subflooring and wall and roof sheathing. Also good for a broad range of other construction and industrial applications. Can be manufactured as conventional veneered plywood, as a composite, or as a nonveneer panel. For special engineered applications, veneered panels conforming to PS 1 may be required. EXPOSURE DURABILITY CLASSIFICATIONS: Exterior, Exposure 1, Exposure 2. COMMON THICKNESSES: 5/16, 3/8, 7/16, 15/32, 1/2, 19/32, 5/8, 23/32, 3/4.

Specially designed as combination subfloor-underlayment. Provides smooth surface for application of carpet and possesses high concentrated and impact load resistance. Can be manufactured as conventional veneered plywood, as a composite, or as a nonveneer panel. 1⅛" plywood panels marked PS 1 may be used for heavy timber roof construction. Available square edge or tongue-and-groove. EXPOSURE DURABILITY CLASSIFICATIONS: Exterior, Exposure 1, Exposure 2. COMMON THICKNESSES: 19/16, 5/8, 23/32, 3/4, 1-⅛.

FIGURE 9. Typical trademarks you may find on plywood and an explanation of what the numbers and terms mean.
American Plywood Association, *APA Design/Construction Guide, Residential & Commercial,* Page 6/7, September 1987

Table III
Veneer Grades

Symbol	Description
A	Smooth, paintable. Not more than 18 neatly made repairs—boat, sled, or router type, and parallel to grain—permitted. May be used for natural finish in less demanding applications. Synthetic repairs permitted.
B	Solid surface. Shims, circular repair plugs, and tight knots to 1 inch across grain permitted. Some minor splits permitted. Synthetic repairs permitted.
C Plugged	Improved C veneer, with splits limited to 1/8 inch width and knotholes and borer holes limited to 1/4 x 1/2 inch. Admits some broken grain. Synthetic repairs permitted.
C	Tight knots to 1-1/2 inch. Knotholes to 1 inch across grain and some to 1-1/2 inch if total width of knots and knotholes is within specified limits. Synthetic or wood repairs. Discoloration and sanding defects that do not impair strength permitted. Limited splits allowed. Stitching permitted.
D	Knots and knotholes to 2-1/2 inch width across grain and 1/2 inch larger within specified limits. Limited splits allowed. Stitching permitted. Limited to Interior, Exposure 1, and Exposure 2 panels.

American Plywood Association, *Grades and Specifications,* Page 5, August 1987.

5. OTHER WOOD PANEL PRODUCTS

Nonveneered or **reconstituted** wood panels are composed of wood particles, flakes, or strands that have been bonded together and formed into sheets.

Technology has progressed to the point that these panels, once only recommended for shelving, interior wall panels, and floor underlayment, can be manufactured strong enough for structural use. These products include waferboard, composite board, structural particleboard, and oriented strand board (Figure 10).

Waferboard. As the name implies, flakes or wood chips called wafers are used in manufacturing waferboard. Wafers are sliced from short logs to specific sizes. The bonding agent is an exterior grade of phenolic resin, which is combined with wafers and hot-pressed into panels typically 4' x 8' and larger. Among other uses, waferboard is used for subflooring, and roof and wall sheathing.

Composite Board. This type of board is made up of a reconstituted wood center, such as particleboard, and faced with a veneer softwood. It is used for underlayment (the materials placed over a subfloor to provide an even, smooth surface for rugs or tile), subflooring, and roof or wall sheathing.

Structural Particleboard. The board is produced by combining very small wood chips or flakes with a resin bonding agent, which is then hot-pressed into panels.

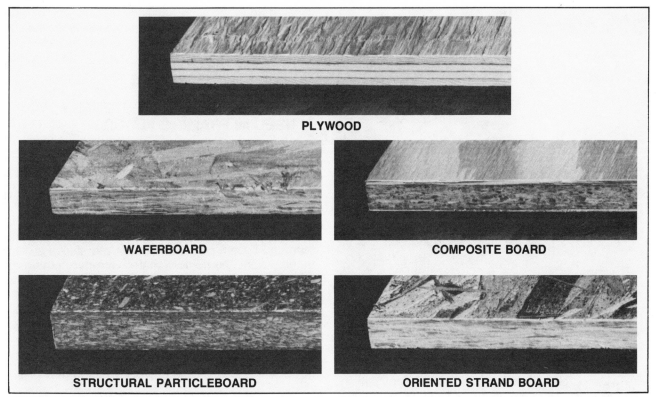

FIGURE 10. Plywood and other, nonveneered or reconstituted wood panels are now used as subflooring materials.

Structural particleboard may also be called chipboard or flakeboard. Until recently, particleboard was used primarily for underlayment and in the construction of cabinets.

Some building codes now list the newer structural particleboard as acceptable for subflooring and roof decking. The main difference between structural and nonstructural particleboard is that structural particleboard has a higher surface density, containing extra resin and wax.

Oriented Strand Board. Oriented strand board is a recently developed structural panel, which is manufactured with strands of wood that are layered perpendicular to each other in three to five layers. A phenolic resin is used to bond the layers into a stiff, smooth, uniform surface. A number of sizes are available in thicknesses ranging from ¼″ to 1⅛″. Oriented strand board can be used for subflooring, wall sheathing, and roof sheathing.

6. CHEMICALLY TREATED LUMBER

Wood is susceptible to serious damage by fungi and termites. Therefore, chemically treated wood may be desirable for some building applications, such as ground or masonry contact and outside exposures.

Damage by some fungi, such as dry rot, breaks down wood tissue and reduces the strength of the wood. Dry rot actually starts with a fungus that lives in a moisture-rich environment, but the damage, or "dry rot," is not evident until the wood has dried out.

Other fungi produce specks, molds, and stains that affect the wood's appearance, but they do not change the structural quality of the wood.

Subterranean termites—small insects that live in underground nests—are among the most damaging to wood that is in direct contact with the ground. These insects make their way to wooden members of a structure via the cracks in foundation walls and through tunnels over masonry.

The use of chemically treated wood eliminates moisture problems and is an effective preventative against damage by insects and fungi.

Although not naturally decay- or termite-resistant, pine, fir, and spruce woods are widely used because they are very receptive to treatment with preservative chemicals. Two of the most commonly used chemical preservatives are ammoniacal copper arsenate (ACA) and chromated copper arsenate (CCA).

By far the most effective way of treating wood is by **pressure-treating.** Pressure-treating involves placing wood into a sealed tank where pressure is used to drive the preserving chemicals into the wood (Figure 11). The result is a wood that insects cannot destroy and that will not rot.

In some areas, treated lumber is being used for both below- and aboveground foundation systems, called permanent wood foundations (PWF). Lumber treated and grade stamped for below-ground usage should be used in constructing wooden foundations.

13

FIGURE 11. Pressure treatment of lumber takes place in sealed tanks, where chemicals are forced into the wood.

When you purchase treated wood for use in floor framing, take notice of the grade stamp or label for information about its intended use (Figure 12). For additional information on pressure-treated wood for PWFs or other applications, contact *American Wood Preservers Association,* P.O. Box 849, Stevensville, Maryland 21666.

Care should be taken when working with treated wood, since exposure to the chemicals used in the treatment process may present certain hazards. For example, treated wood should not be burned in open fires, stoves or fireplaces because toxic chemicals may be produced as part of the smoke and ashes. After working with treated wood, exposed areas of your skin should be washed before eating or drinking, and prolonged inhalation of sawdust from treated wood should be avoided.

Treated wood should be handled in accordance with the recommendations of the agency responsible for treating the lumber. Consumer information pamphlets should be available upon request wherever treated wood is sold.

7. NAILS AND OTHER FASTENERS USED IN FRAMING FLOORS

Although there are many different kinds of metal fastening devices, nails are still the most widely used. In our discussion, we·will focus on nails and staples, the most commonly used fasteners in floor framing.

Nails. Nails are made of many materials, such as aluminum and coated cement or galvanized steel. For floor framing work, the cement-coated wire nail is the most widely used (Figure 13). It is cut from wire, given a head and a point, and then coated with an adhesive to provide increased holding power. Common wire nails are available in sizes ranging from 2d (1″ long) to 60d (6″ long). See Figure 14 for an example and sizes.

When working with treated lumber, hot-dipped, galvanized nails should be used. Nails with spiral shanks pro-

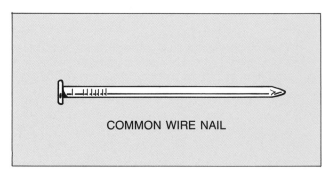

COMMON WIRE NAIL

FIGURE 13. Cement-coated wire nails are commonly used for framing work.

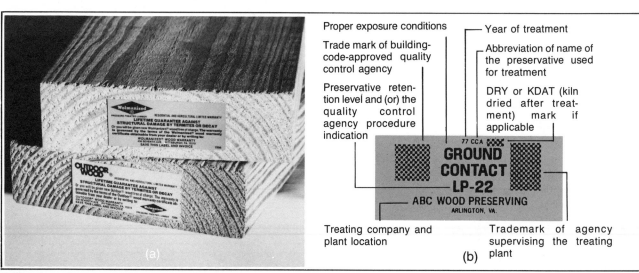

Proper exposure conditions

Trade mark of building-code-approved quality control agency

Preservative retention level and (or) the quality control agency procedure indication

Year of treatment

Abbreviation of name of the preservative used for treatment

DRY or KDAT (kiln dried after treatment) mark if applicable

77 CCA

GROUND CONTACT LP-22

ABC WOOD PRESERVING
ARLINGTON, VA.

Treating company and plant location

Trademark of agency supervising the treating plant

(a)

(b)

FIGURE 12. The (a) label or (b) grade stamp found on pressure-treated wood will provide information about its intended use.

(a) Koppers Company, Inc., (b) Western Wood Products Association, *Treated Lumber,* November 1987.

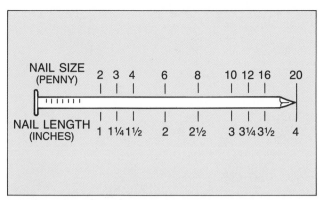

FIGURE 14. A comparison between so-called "penny" sizes of nails and their actual length in inches.

vide extra holding power that will aid in keeping warp-prone treated lumber fastened straight.

Nail sizes are designated by a number and the letter **d.** The letter **d** stands for **denarius**—the Roman word for coin or penny. Sizes of nails are still referred to by the term penny, hailing back to the time when they were purchased in England according to how many pennies they cost per hundred. Larger nails cost more per hundred than do smaller sizes. Today, nails are sold by the pound. Builders generally buy nails by the 50 lb. box.

Staples. Staples are now used by some builders for functions where nails would have been used in the past. They are now generally used to fasten subflooring. Heavy-duty staples are driven in by electric or pneumatically driven tools. Staples come in a variety of shapes and sizes.

Screws. With the availability of automatic self-feeding screwdrivers, the use of screws to secure subflooring is increasing. The screws are packaged in a special plastic belt and are fed into a drive position (Figure 15).

Using screws, combined with a special adhesive applied to the joists, helps eliminate squeaking floors.

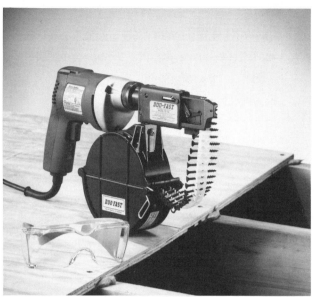

FIGURE 15. Using screws, installed by a self-feeding screwdriver, helps eliminate squeaking subflooring.

15

STUDY QUESTIONS

1. The primary material used for floor framing is _____.

2. _____ grades of plywood are acceptable for use in subflooring or decking.

3. What are the two methods of drying lumber?

4. What do the letters "KD" on lumber represent?

5. Lumber is graded after it has been _____.

6. Softwood lumber is size classified as _____, _____, and _____.

7. Lumber used in floor framing is mainly in the _____ category.

8. Pieces of lumber up to 2″ thick are called _____.

9. Pieces of wood 5″ thick and above are called _____.

10. The _____ dimensions are the measurements we use when we refer to a piece of lumber's thickness and width.

11. Lumber is figured and sold based on its _____ dimensions.

12. The true measure of a piece of lumber is its _____ measure.

13. Lumber measurements are given as _____, _____, and _____.

14. The _____ will always be actual size.

15. A board foot of lumber is equal to any piece of lumber with a volume of _____.

16. Give the actual size of:
(a) 1″ x 4″ _____.
(b) 2″ x 6″ _____.
(c) 4″ x 4″ _____.

17. Fill in the values used in the lumber board feet formula

$$\frac{NO.\_\_\_\_ \times T\_\_\_\_ \times W\_\_\_\_ \times L\_\_\_\_}{12}$$

18. How many board feet in the following lumber bill: 2 pieces 2″x10″x12′, and 4 pieces 1″x6″x10′. Total _____ B.F.

19. _____ has almost replaced lumber as a subflooring material.

20. Sheets of wood veneer are glued together by a combination of _____ and _____.

21. A standard size sheet of plywood measures _____ (width) x _____ (length).

22. Name three other wood panel products _____, _____, and _____.

23. Lumber used in ground contact, masonry contact and outside (e.g. decks) should be _____.

24. The most commonly used fastener in floor framing is the _____.

25. The letter "d" when referring to nails stands for denarius, which means _____.

After the foundation work has been completed, the next step is floor framing. Two types of floor framing used in the construction of dwellings will be discussed:

1. Platform Framing
2. Two-Story Framing

1. PLATFORM FRAMING

Platform framing is now used for most residential construction. The floor frame is constructed on the foundation wall and concrete block piers in a platform manner.

House plans contain many terms that refer to the parts of a house. You will become familiar with most of these as you gain experience. However, there are several basic structural parts you should be able to identify from the start. Knowing the names of the parts will help you understand framing floors more clearly and will enable you to communicate with others about the job.

The floor frame is made up of the following components (Figure 16).

Footing—base of the foundation system which is in direct contact with the soil.

Foundation—masonry material between the footing and lumber of the structure. It provides support for the loads above it.

Sill, sill plate, bond plate, or mudsill—a piece of dimensional lumber that is fastened to the top of a foundation wall. This plate is the nailing base for floor joists or studs. It bonds and anchors the wood frame to the foundation.

Floor joists—horizontal planks placed on edge upon which the subfloor will be nailed.

Header joists—pieces of lumber the same size as the floor joists. They are nailed to the ends of the floor joists in a continuous band to prevent the floor joists from tipping or rolling.

End joists—horizontal planks at either end of the floor frame running in the same direction as the floor joists.

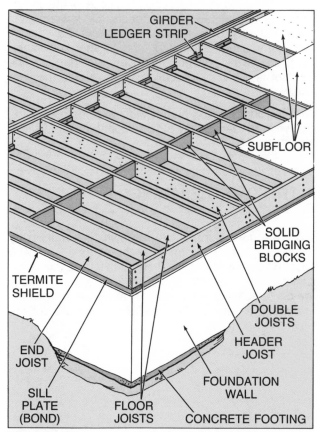

FIGURE 16. Components of a basic floor frame.

Double joists—two joists nailed together to provide additional strength under a parallel wall. Blocks of wood may be placed between doubled joists to allow for water pipes, heat ducts, or vent stacks coming from a basement or underneath the house.

Girder—a large horizontal member made of steel and/or of wood, used to help support a floor, ceiling, or roof. It carries one half the load of the floor joists.

Ledger or ledger strip—a strip of lumber nailed at the bottom of a girder, designed to give support to joists butting against the girder.

Bridging (joists)—wood or metal pieces nailed or fastened between the joists to hold them in position. There are two types of wooden bridging—solid and herringbone cross. Bridging also helps to distribute dead and live loads.

Note: The *dead load* is the weight of the structure of a building which includes all the materials that make up the floors, walls, ceilings, roofs, etc.

The *live load* includes all the moving and changing loads placed on the different sections of the house. Live loads include people, furniture, and factors such as wind and snow.

Subfloor—a structure of boards or panels which is fastened to the floor frame. It provides a base for the underlayment and finished floor materials. (Underlayment is material placed over the subfloor to provide a smooth, even surface for the carpet or other finished floor material. It may be thin plywood or nonveneered panels). The subfloor is the first application of material to the floor frame.

After the floor framing and subfloor are completed, the carpenter is ready to lay out and construct the walls of the house.

2. TWO-STORY FRAMING

When a second story is built, the downstairs walls of the first floor serve as supports for the second floor. The second floor frame is constructed similarly to that of the first floor. Each platform is built separately, and the walls are constructed after the subfloor materials are placed. The walls of the first floor serve the purpose for the

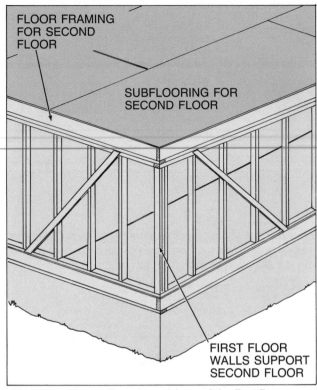

FLOOR FRAMING FOR SECOND FLOOR

SUBFLOORING FOR SECOND FLOOR

FIRST FLOOR WALLS SUPPORT SECOND FLOOR

FIGURE 17. The walls and partitions of the first floor serve as foundation for the second floor in two-story houses.

second story that the foundation and piers (and girder) serve for the first floor. (Figure 17).

NOTE: To support joist ends, it is often necessary to construct a load-bearing girder when framing the second floor.

STUDY QUESTIONS

1. Name the two types of floor framing discussed in this unit.

2. Give two reasons for learning the names of floor parts.

3. The _____ is the first piece of lumber placed in floor framing.

4. The subfloor will be nailed to the _____.

5. _____ are nailed to the end of floor joists.

6. A _____ supports the interior end of floor joists.

7. Name two types of bridging.

8. The _____ provides a base for the underlayment and finished floor.

9. State two purposes of bridging blocks.

10. On two-story construction, the _____ support the upstairs platform.

There are a number of factors that influence the materials used to frame floors:
• Architect's design
• Specifications
• Availability of materials
• Comparative costs
• Builder's preference
• Local codes

Architect's design. The design of a house may require the use of special lumber. For example, longer floor joists to span large open areas may be required. In some cases, a design calling for exposed beams in a basement or other area will necessitate that large solid or laminated timbers be ordered.

Specifications. When a complete set of building plans is purchased, generally a document called the **specifications** or **specs** is included. Sometimes these specifications may be broken into sections or divisions covering specific areas of the total job.

In one of these sections, you may find information connected with floor framing including such things as the grade and species of lumber for framing, and the thickness and the APA grade of plywood to be used. Some specifications include a materials listing.

Availability of materials. For a number of reasons, certain building materials may be unavailable at the time you need them. When shortages do occur, the builder must be knowledgeable enough to select substitute materials from among those that are available. The substitutes should be structurally as strong as those originally specified, as well as cost compatible.

Comparative costs. Alternate materials may be available at a substantial savings over those that are inconvenient to locate or unreasonably priced. **Again, the builder must be knowledgeable in the area of material compatibility to make sure that unsound or inferior materials are not used.**

Builder's preference. A builder may choose to use certain materials because of personal factors such as his familiarity with the characteristics and workability of specific lumber species.

Local codes. Codes in some areas of the country specify minimum standards for lumber used in the construction of floor frames.

Selecting and estimating flooring materials is discussed under the following headings:

1. Selecting Materials for Floor Framing
2. Estimating Lumber for Floor Framing
3. Selecting Materials for Subfloors
4. Estimating Lumber for Subfloors
5. Estimating Plywood or Other Panel Products for Subfloors

1. SELECTING MATERIALS FOR FLOOR FRAMING

The selection of lumber to be used in framing a floor will be dependent upon a number of factors. Some of these factors like local code requirements, architect's design, specifications, price, and availability of some materials have already been discussed.

Other factors to be considered are as follows:

• Span
• Load
• On-center spacing
• Size of lumber
• Species of lumber

Span. The distance between the foundation wall and the center girder or support pier is the span. The longer the span, the larger and stronger the lumber used must be. Table IV provides information on the *allowable spans* for various grades and sizes of lumber commonly used for floor joists. The allowable span is the maximum distance allowed between supporting points for given sizes of lumber.

Load. The meanings of live and dead loads have already been discussed. Table IV gives information on selecting joists size based on the average live and dead loads for residential living areas.

19

Table IV

Allowable Spans — Floor Joists
(Living Areas)

Liveload 40 lb./sq. ft.
Deadload 10 lb./sq. ft.
Deflection — 1/360th of span

Spans Are Based on Repetitive Member
Values for Normal Duration of Loading

| Size in Inches | Spacing in Inches | Maximum Allowable Spans in Feet and Inches | | | | | | | | | | |
| | | Southern Pine | | | | | Douglas Fir-Larch | | Hem-Fir | | Spruce-Pine-Fur | |
		No. 1 KD*	No. 1 SD	No. 2 KD*	No. 2 SD	No. 3 SD	No. 2	No. 3	No. 2	No. 3	No. 2	No. 3
2x6	12	11- 2	10-11	10- 9	10- 9	9- 0	10-10	9- 3	10- 3	8- 3	10- 0	7- 7
	16	10- 2	9-11	9- 9	7- 9	9-11	8- 0	9- 4	7- 2	8- 8	6- 7	5- 4
	24	8-10	8- 8	8- 6	8- 5	6- 4	8- 6	6- 5	7- 7	5-10	7- 1	5- 4
2x8	12	14- 8	14- 5	14- 2	14- 2	11-10	14- 4	12- 2	13- 6	10-11	13- 2	10- 0
	16	13- 4	13- 1	12-10	12-10	10- 3	13- 1	10- 7	12- 3	9- 5	11- 6	8- 8
	24	11- 8	11- 5	11- 3	11- 1	8- 4	11- 3	8- 7	10- 0	7- 8	9- 4	7- 1
2x10	12	18- 9	18- 5	18- 0	18- 0	14- 1	18- 4	14- 7	17- 3	13-11	16-10	12-10
	16	17- 0	16- 9	16- 5	16- 5	13- 1	16- 8	13- 6	15- 8	12- 0	14- 8	11- 1
	24	14-11	14- 7	14- 4	14- 2	10- 8	14- 4	11- 0	12-10	9-10	11-11	9- 1
2x12	12	22-10	22- 5	21-11	21-11	18- 4	22- 4	18-10	21- 0	16-11	20- 6	15- 7
	16	20- 9	20- 4	19-11	19-11	15-11	20- 3	16- 4	19- 1	14- 7	17- 9	13- 6
	24	18- 1	17- 9	17- 5	17- 2	13- 0	17- 6	13- 4	15- 7	11-11	14- 6	11- 0

*KD — (Surfaced at 15 percent maximum moisture content-KD)
The allowable span is the clear distance between supports.

On-center spacing. Also referred to as "O.C.", the on-center spacing is the distance from the center of one supporting member to the center of the next, usually 12", 16", or 24". The greater the anticipated load or the longer the span the closer the spacing.

Size of lumber. The larger the lumber is physically, the stronger it is. Table IV lists the sizes of floor joists commonly used. The size used will be determined by a combination of the factors being discussed.

Species of lumber. Certain species of lumber are stronger structurally than others of the same size. Table IV lists the wood species commonly used for joists. Use of some wood may require shorter spans or closer spacing between joists.

Selecting materials for floor framing is discussed under the following headings:

a. Selecting Materials for Sills and Joists
b. Selecting Materials for Ledger Strips and Bridging

a. Selecting Materials for Sills and Joists

Generally, the lumber size for sills and joists will be either 2" x 6", 2" x 8", 2" x 10", or 2" x 12" (nominal sizes). No. 2 grade is often specified. Utility grade and any grade in any material under No. 3 are seldom permissible for use in floor framing. Remember that lumber is available in lengths of 8' - 24' in increments of 2', with the longer lengths costing more.

b. Selecting Materials for Ledger Strips and Bridging.

Lumber selected for ledger strips should be of the same grade and species as the floor joists. The sizes used will be 2" x 2" or 2" x 4" (nominal sizes). Since the ledger strips will be fastened to the bottom edge of a girder for the support of joists, overall length is not important. Piecing the ledger strip will not weaken its strength for the purpose for which it is designed.

There are two types of wooden bridging—solid and herringbone cross.

Solid bridging utilizes lumber of the same dimensions as the floor joists (Figure 18). Since these blocks will be cut to fit between the joists, the beginning length is not important. The blocks are staggered to make nailing them in place easier.

Herringbone cross bridging is made up of a pair of narrow wooden or metal pieces set diagonally between floor joists (Figure 18). For this type of bridging, you use 1" x 4" boards. Again, since the bridging members will be cut to fit, the beginning length of the lumber is not important.

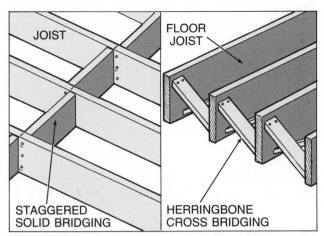

FIGURE 18. The two types of wooden bridging are (a) staggered solid block bridging and (b) herringbone cross bridging.

2. ESTIMATING LUMBER FOR FLOOR FRAMING

Floor joists. The number of pieces of lumber needed for a given floor structure is fairly easy to figure. Since the most common spacing for floor joists is 16″ O.C. (on-center), you may use it as an example. **Remember that extra and double joists are required under all partitions that run in the same directions as the joists.**

As an example, calculate the lumber required in a floor for a house 28′ x 60′. For purposes of this example, assume that you will use one center girder. By use of a diagram, you can see that two runs of 14′ joists will be needed to span the 28′ width of the house (Figure 19).

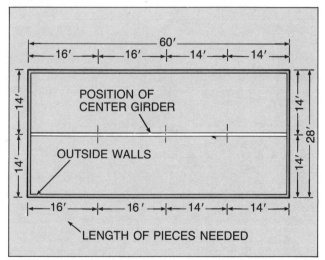

FIGURE 19. A simple diagram will help you determine the lumber length needed for sills and girders.

Now that you know the length of lumber needed for floor joists, you must determine how many will be needed. Table IV provides information showing you that if you have decided to place your joists 16″ O.C., you'll need to use 2″ x 10″ No. 2 Southern Pine or Western Wood to span 14′.

Multiply the length of the house (run) in feet by the number of inches in a foot (12″) to determine the total inches in each run:

60′ x 12″ = 720″ total inches in each run

Divide the total (720″) by the distance between the joists (16″):

720 ÷ 16″ = 45 joists needed for one run

Add 1 for a "starter joist" = 46 joists (one run). Since two runs of 14′ lumber will be required, multiply the number of joists needed for one run (46) by 2:

46 x 2 = 92 joists needed

Add to this any double or extra joists needed under **parallel walls.** Assume for the purposes of this example that there are eight such walls:

92 + 8 = 100 joists 2″ x 10″ x 14′ needed

Bridging. When joists span 8′ to 15′ there should be one run of bridging installed half way between the girder and the supporting wall. Since there are two runs of joists spanning 14′ each, two 60′ rows of bridging (the length of the house) will be required:

60′ x 2′ = 120′ (two runs)

Divide the length of lumber being used (14′) into the total length of the two runs (120′):

120 ÷ 14 = 8.57 pieces 14′ long needed

Remember that the lumber will be cut into short lengths for bridging.

Headers. Although lumber is obtainable in longer lengths, consider lengths of 12′, 14′, and 16′ when preparing a materials list since these lengths are more readily available. By referring to the diagram in Figure 19, you can see that the use of two 16′ and two 14′ pieces will be needed to construct a header down each side. Since you'll need material for two headersl, you should add the following to your materials list:

4 — 2″ x 10″ x 16′
4 — 2″ x 10″ x 14′

So far the material list should look this way:

Material	Purpose	Number Needed
2″x10″x14′	in line joists	92
2″x10′x14′	extra and doubles	8
2″x10″x14′	solid bridging and part of headers	13
Total number 2″ x 10″ x 14′ needed		**113**
2″ x 10″ x 16′	part of header	4
Total number 2″ x 10″ x 16′ needed		**4**

Materials for sill plates and the girder must also be added.

Sill plates. Sill plates are usually made of 2″x6″ or 2″x8″ lumber. To conserve lumber and save money, use pieces in lengths that will minimize waste. The diagram in Figure 19 indicates that using 8 pieces 14′ long and 4 pieces 16′ long will provide enough lumber to construct the sill plate (176′ in our example). Add to your materials list:

Material	Purpose	Number Needed
2″x6″x14″	sill plate	8
2″x6″x16″	sill plate	4

Girders. The girder runs the length of the house (60′ in our example). It can be a double or triple girder. Either 2″ x 10″ or 2″ x 12″ lumber is usually used for the girder. For a more detailed discussion of building girders and ledger strips, see Section F-2, *Building and Installing a Butt-to-Girder.*

As you can see by looking at the diagram in Figure 19, one 60′ run will take:

$$2 - 2″ \times 10″ \times 16′$$
$$2 - 2″ \times 10″ \times 14′$$

The figures must be doubled for a double girder, add to your materials list:

Material	Purpose	Number Needed
2″ x 10″ x 14′	girder	4
2″ x 10″ x 16′	girder	4

If you were constructing a triple girder, the number needed would be 6, not 4, of each.

Nailed to the bottom edge of each side of the 60′ girder (double or triple) will be a ledger strip. You will need a total of 120 linear feet of 2″ x 2″, which can be ripped from 60′ of 2″ x 4″ pieces. Add to your material list:

2 — 2″ x 4″ x 16′
2 — 2″ x 4″ x 14′ = 60′ of 2″ x 4″ (which when ripped to 2″x2″ strips, will produce 120 linear feet)

Add these figures to the materials list, which should look like this:

Material	Purpose	Number Needed
2″x10″x14′	in-line joists, extras and doubles, headers, and solid bridging	113
2″x6″x14′	sill plates	8
2″x6″x16′	sill plates	4
2″x10″x16′	double girder	4
2″x10″x14′	double girder	4
2″x4″x16′	ledger strip	2
2″x4″x14′	ledger strip	2

NOTE: Calculations do not allow for damaged and defective lumber discovered after delivery and inspection at the site, so ordering a bit extra, say 10%, is a good idea.

Materials for any special openings will have to be added as needed. Individual examples will be given later in this publication.

Other materials will have to be added to the materials list. Among these are nails and the metal to be used as termite shields for foundation walls and piers (described later).

---

3. SELECTING MATERIALS FOR SUBFLOORS

Lumber, plywood or an approved panel product serves well as subfloor material (Figure 20).

When lumber is used, it should be 1″ x 6″ or 1″ x 8″ (minimum ¾″ thick if surfaced on both sides). It should run on a 45° (degree) diagonal across the floor joists (Figure 20). In most homes built today, plywood or other panel products are the materials used for subfloors. Plywood products are highly acceptable, easily installed, strong, and when used according to recommendations of the manufacturer, present few problems. One of the most popular panels used for subflooring is 15/32″, 1/2″, or 5/8″ 3- and 4-ply sheathing, called "CDX" in the trade.

Check Figures 9 and 10 for information concerning grades and suggested uses. Plywood sheathing is grade stamped for its particular intended use. This stamp includes a "Span Rating" number that may be used to insure proper installation. For in-depth information on plywood and other panel products, contact the American Plywood Association, P.O. Box 11700, Tacoma, Washington 98411.

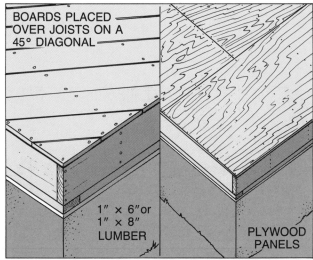

BOARDS PLACED OVER JOISTS ON A 45° DIAGONAL

1″ x 6″ or 1″ x 8″ LUMBER

PLYWOOD PANELS

FIGURE 20. Subflooring may be (a) lumber (boards) or (b) plywood or other panel products.

4. ESTIMATING LUMBER FOR SUBFLOORS

Because the actual width of boards is less than nominal measure and a waste figure has to be added in, 120 percent of floor area should be used in estimating materials to be used.

Using the house plan we've been working with, we have a floor area 60' x 28' or 1,680 square feet (length x width). Multiply the floor area (1,680) x 120 percent to arrive at the total board feet needed.

1,680' x 120% = 1,680 x 1.2 = **2,016 board feet needed**

To determine the number of linear feet of materials needed, use the following procedure:

If using 1"x6" boards—Multiply 2 x number of total board feet needed since 6" x 2 = a foot (12").

2 x 2,016 = **4,032 linear feet of 1" x 6"**

If using 1" x 8" boards—Multiply 1.5 x number of total board feet since 8" x 1.5 = 12".

1.5 x 2,016 = **3,024 linear feet of 1" x 8"**

5. ESTIMATING PLYWOOD OR OTHER PANEL PRODUCTS FOR SUBFLOORS

When plywood is selected, grade "CDX", 3- or 4-ply, 15/32", 1/2" or 5/8" thick panels are most often used. Plywood (and most panel products) are available in 4' x 8' sheets (4' x 8' = 32 square feet).

The number of pieces needed can be determined by first determining the square feet in the framing platform to be covered. For the 60' x 28' house in our example, we determined that the floor area was 1,680 square feet (length x width). Divide the number of square feet in the house (1,680) by the number of square feet in a sheet of plywood (32):

$$\frac{1,680}{32} = 52.5 \text{ or } 53 \text{ pieces}$$

Sometimes an odd-sized or odd-shaped floor will require additional sheets or pieces, so you should evaluate your floor plan when estimating material.

Although plywood is the oldest and most widely accepted material for subfloors, the use of nonveneered panels, such as structural particleboard and others, is growing. All panels used for the subfloor must be of the proper grade and thickness for the requirements of your local building code.

STUDY QUESTIONS

1. List four things that influence the material used for floor framing:

2. The species of lumber generally used in framing floors is _____ or _____.

3. The letters "O.C." on a floor plan mean _____.

4. The lumber grade is usually specified for floor framing is _____.

5. Define the term **joist span.**

6. Herringbone cross bridging is usually made from what size of boards?

7. What is the most common on center spacing for floor joists?

8. Code requires double joists under all _____.

9. Sill plates are usually made of _____ x _____ or _____ x _____ lumber.

10. Plywood _____ inch thick in grade _____ is often used for subflooring.

The sill plates are the first pieces of the floor frame to be assembled. They are fastened to the top of the foundation walls. To protect the sill plate from possible infestation by termites, a barrier must also be installed. Termite protection is required by code in many states.

Installing termite shields and sill plates will be discussed under the following headings:

1. Installing Termite Shields
2. Installing Sill Plates

1. INSTALLING TERMITE SHIELDS

Termites cause millions of dollars in damage to homes every year. They build tunnels from the earth to provide a route to the wood structure of a home (Figure 21). Only a few states in the driest and coldest parts of the U.S. escape the threat of damage. Many lending institutions in areas subject to termite attack require a certificate of treatment and other preventative measures before financing will be provided.

FIGURE 21. Termites can do tremendous undetected damage unless preventative measures are taken.

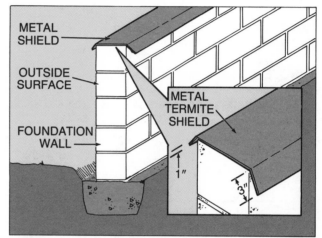

FIGURE 22. Correct placement of a metal termite shield over a foundation wall.

Two methods for termite protection are soil treatment and termite shields. Since soil treatment takes place in conjunction with the foundation, only termite shields will be discussed here.

A thin sheet of metal is positioned over the foundation wall before the sill plate is placed and secured. This metal sheet helps keep termites away from the wood members of the house. It is known as a termite shield. The termites cannot tunnel through or climb over the metal.

When a termite shield is not used (e.g. in areas where termites are not a problem), a layer of asphalt-saturated felt, placed between the foundation wall and the wood sill, may be used to keep moisture from coming into direct contact with the sills.

To install sheet metal for termite protection, proceed as follows:

1. *Use a sheet wide enough to cover the foundation wall and bend down over the sides.*

   Bend the metal over the outside wall about 1″ and over the inside wall about 3″ (12″ metal works well over an 8″ block wall, as shown in Figure 22).

25

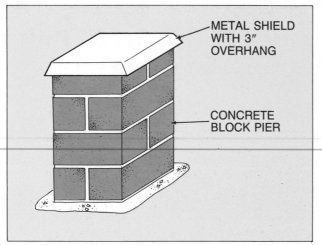

METAL SHIELD WITH 3" OVERHANG

CONCRETE BLOCK PIER

FIGURE 23. Metal termite shields should be placed over the piers as well as the foundation walls.

NOTE: If anchor bolts are used, you must punch or drill holes in the material to allow the shield to fit next to the foundation wall. Holes punched into the shield material should be sealed around the bolts.

2. *Place a shield over any piers and around the foundation walls of fireplaces and pipes (Figure 23). Allow 2" to 3" to overhang.*

2. INSTALLING SILL PLATES

Sill plates are not only the first pieces of the floor frame, they are also the base upon which the header joists and one end of the floor joists will rest.

To install a sill plate proceed as follows:

1. *Place insulation on the foundation wall, atop the termite shield.*

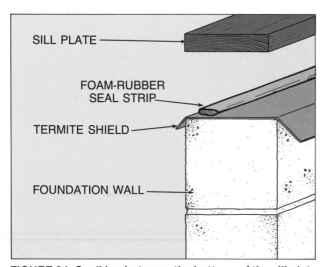

SILL PLATE

FOAM-RUBBER SEAL STRIP

TERMITE SHIELD

FOUNDATION WALL

FIGURE 24. Caulking between the bottom of the sill plate and the top of the termite shield will help prevent air infiltration.

26

To prevent air infiltration, install caulking or a foam-rubber seal strip between the termite shield and the sill plate (Figure 24). Caulking may be done later, before the outside wall is finished, by sealing along the metal and wood joint.

2. *Lay the sill stock along the top of the foundation walls as near as possible to the location where they will be secured. Do not join over foundation openings or other openings, such as scuttle openings or ventilation holes.*

Check your local building codes to determine if treated lumber is required for sills. The material used may be either 2" x 6" or 2" x 8" stock (see Section C-2, *Estimating Lumber for Floor Framing*).

The sill plate may be positioned on the foundation wall so that its outer edge is flush with the outer face of the masonry. This allows the wall sheathing to extend down the outside of the wall below the sill plate (Figure 25a).

Some builders like to leave a sheathing ledge on the foundation wall and will backset the sill plate by the thickness of the sheathing to be used (Figure 25b).

Another technique is to *set in* the frame wall on top of the subfloor, allowing space to install sheathing (Figure 25C).

When this last procedure is followed, it is wise to seal the floor and wall bottom with a strip of 30# felt. The felt will act as a barrier against air infiltration at this point. An 18" strip of felt, stapled to the bottom of the studs and floor framing before the finished wall material is installed, works ideally.

3. *Check sill for leveling.*

It is important that the sill plates be level. If the foundation wall is not level, shims (thin, wedge-shaped strips) must be placed where needed to make the sill level (Figure 26).

Installation of a level footing in the first place will help save extra time and effort later and help to carry that level throughout the building process.

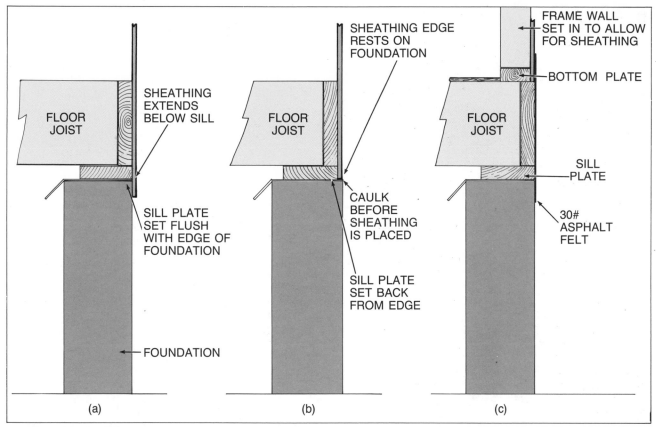

FIGURE 25. Sill plates may be (a) positioned on the edge of the foundation wall with sheathing extending below the sill; (b) set back by the thickness of the sheathing; or (c) set on the edge with the frame wall set in enough to allow for installation of the sheathing.

4. *Fasten sills with anchor bolts.*

Anchor bolts provide the best system of securing a sill plate to the foundation wall (Figure 27). When anchor bolts are used, they should be installed by the mason who builds the foundation. These bolts should be installed in compliance with building code recommendations.

Some builders prefer to nail the sills to the masonry using steel cut nails. If sills are to be nailed, a solid concrete block should be used on top of the foundation wall, since the web of a hollow line block provides little space to drive the nails.

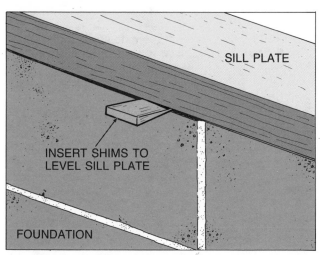

FIGURE 26. Installation of shims may be necessary to level the sill if the foundation wall is not level.

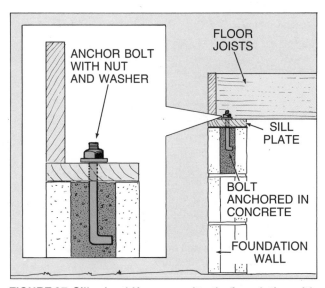

FIGURE 27. Sills should be secured to the foundation with anchor bolts.

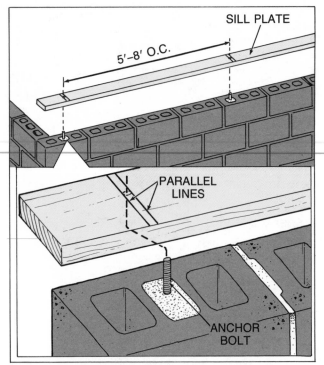

FIGURE 28. Draw parallel lines across the sill plate at the proposed bolt location. The termite shield is not illustrated so that bolt locations can be shown more clearly.

For purposes of this discussion, it is assumed that anchor bolts have been placed in the foundation wall.

Proceed as follows:

(a) *Place sill stock along the edge of the foundation.*

(b) *Draw parallel lines across the sill plate at the bolt location* (Figure 28).

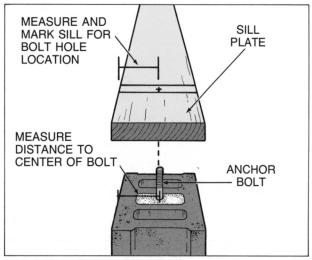

FIGURE 29. Measure and mark the distance from the outside edge to the bolt location.

(c) *At parallel lines measure the correct distance from the outside edge of the sill to the center line of the bolt location* (Figure 29).

Keep in mind the desired position of the sill, remembering the earlier discussion concerning backsetting of the sill.

(d) *Using a wood bore bit, drill anchor holes ¼" larger than the bolt diameter.*

Using larger holes will make the fitting and alignment of the sill plate easier.

Mark and cut square the sill stock allowed to run over the corners of foundation wall.

5. *With the termite shield in position, place the sill plate over the anchor bolts (Figure 30a).*

6. *Place washers and nuts on the anchor bolts and tighten until snug (Figure 30b).*

Do not overtighten the bolt, as this may draw the sill down too much or cause the bolt to break out of the masonry.

7. *Toenail the ends of the sill plates to each other.*

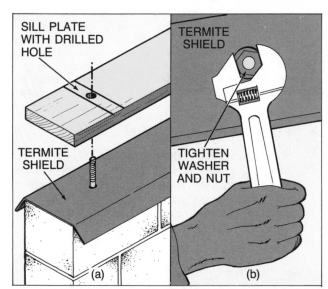

FIGURE 30. (a) Position the termite shield and sill plate over the bolt and (b) place washer and nut on bolt and tighten.

STUDY QUESTIONS

1. Name the first piece of floor framing assembled.

2. What are the two methods of providing termite protection.

3. A thin metal sheet is positioned over the _____ before the sill plate is installed.

4. The termite shield should extend over the foundation wall _____ to the outside and _____ to the inside.

5. True or false. Shield metal is not necessary over piers.

6. The header joists and one end of the floor joists rest upon the _____.

7. To prevent air infiltration, install _____ or _____ between the termite shield and the sill plate.

8. _____ may be used when necessary to level the sill plate on the foundation.

9. The best system of securing sills to the foundation wall is the use of _____.

10. Briefly list five steps for installing sill plates with anchor bolts:

NOTES

After the sill plates have been secured to the foundation wall, the next step is the building of the perimeter sill (also known as the box sill).

The sill plates and the perimeter sill will form the outline of the floor structure and will look somewhat like a box (Figure 31).

The perimeter sill is made up of end joists (parallel to the floor joists) at each end of the house and header joists at the end of the floor joists. You need to be careful to fit the various pieces together flush (even) and neat.

To build the perimenter sill proceed as follows:

1. *Measure and mark end joists and header joists that make up the perimeter sill.*

These pieces of lumber will be the same size stock as the floor joist material to be used (2" x 8", 2" x 10", or 2" x 12").

2. *Square lumber ends and saw lumber to fit.*

Perimeter sill lumber that has to be pieced together should be joined by toe nailing or by the use of a **cleat.** A cleat is a narrow piece of wood spanning the joint between the two edges which is nailed to both boards.

3. *Secure the header joists at the corners (where they overlap the end joists) by nailing together with three 16d nails if using 2" x 6" or 2" x 8" lumber or four 16d to 20d nails if using 2" x 10" or 2" x 12" lumber (Figure 32).*

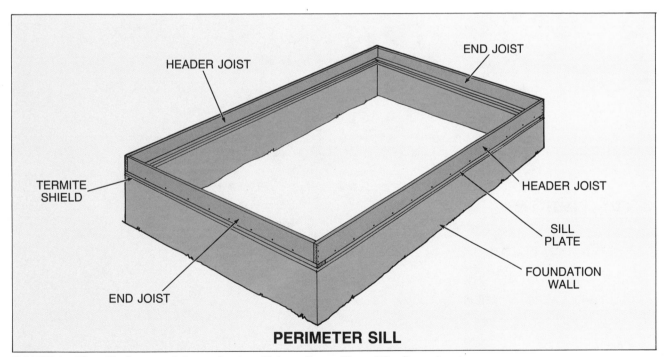

PERIMETER SILL

FIGURE 31. The sill plates and the perimeter sill form the outline of the floor structure.

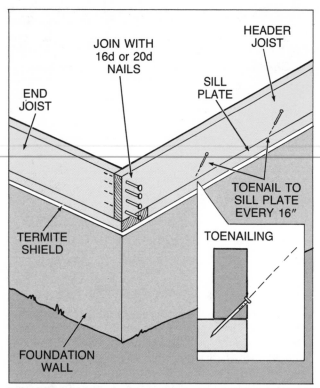

END JOIST

JOIN WITH 16d or 20d NAILS

HEADER JOIST

SILL PLATE

TERMITE SHIELD

TOENAIL TO SILL PLATE EVERY 16″

TOENAILING

FOUNDATION WALL

FIGURE 32. Nail the header joists to the end joists with 16d to 20d nails.

4. *Secure the header and end joists to the box sill plate by toenailing them with 12d or 16d nails every 16 inches (Figure 32).*

NOTE: Toenailing is the driving of nails at a slant into two pieces of lumber.

STUDY QUESTIONS

1. The perimenter sill is made up of _____ joists and _____ joists.

2. Floor pieces should fit together _____ and _____.

3. Header joists are made from the same size lumber as the _____.

4. _____ (size) nails are used to secure 2″ x 10″ header joists to end joists.

5. Driving of nails at a slant into two pieces of lumber is called _____.

F.   Constructing and Installing Built-Up Wood Girders

Once the perimeter sill is secured to the foundation, constructing a girder is the next step in building a floor frame. There are two classifications of girders—**non-bearing** and **bearing.**

Nonbearing girders support the **dead** and **live** loads of the floor system. Remember, the dead load is the combined weight of the material used to construct the floor. The live load is the weight to be supported, made up of people and items placed on the floor when the structure is occupied. Bearing girders not only support the dead and live loads, they must also support a wall framed directly above.

Girders are designed to hold up the lapped and butted ends of the interior floor joists. They provide central support for longer joist spans. Their size depends upon the load they are designed to carry (see Table IV). Girders rest upon piers, columns, or posts.

Girders may be constructed on the job by nailing three pieces of lumber together, or they may be bought as laminated girders in which the timbers are glued together. Steel beams may be purchased and used in conjunction with wood framing. The joists either rest upon the top of the steel beam or butt against the sides.

The following discussion gives procedures for building wooden girders on-site. A number of methods exist for the construction and placement of girders. Two basic methods are discussed under the following headings:

1. Building and Installing a Dropped Girder

2. Building and Installing a Butt-to-Girder

1. BUILDING AND INSTALLING A DROPPED GIRDER

A girder positioned so as to allow floor joists to lay over it or on top of it is known as a **dropped girder** (Figure 33).

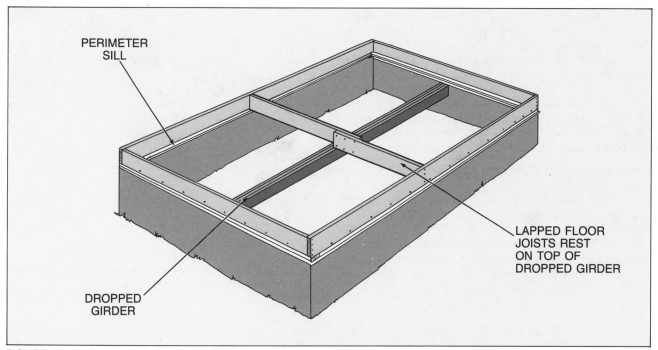

PERIMETER SILL

LAPPED FLOOR JOISTS REST ON TOP OF DROPPED GIRDER

DROPPED GIRDER

FIGURE 33. A dropped girder is positioned so that the bottoms of the floor joists rest on top of the girder.

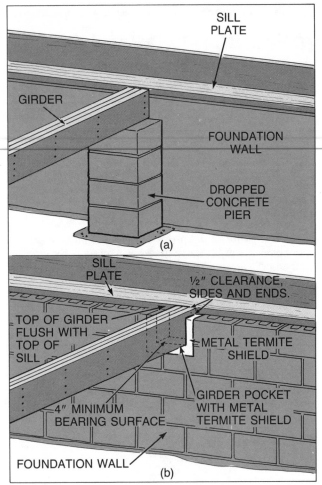

FIGURE 34. The end of a dropped girder may (a) rest on a shortened pier of blocks or concrete or (b) be positioned in a specially provided girder pocket.

The dropped girder is placed so that the top is flush with the bottom of the floor joists and the top of the sill plate. This positioning means that the girder needs support: either a shortened pier at each end to rest upon or **girder pockets** in the foundation (Figure 34). When a girder pocket is used, the pocket must be large enough to allow ½″ air space around the sides and ends of the girder and must be at least deep enough to provide a 4″ minimum bearing surface on the wall. To protect against termites, the pocket should be lined with metal.

Although a dropped girder provides good support and is easy to install, a short distance from the floor framework to the earth underneath (crawl space) sometimes prohibits its use. The minimum crawl space required by most building codes is 18″ to 24″. Without raising the foundation, this distance may be impossible to achieve. The dropped girder can be used in any house with space above the.minimum clearance.

With the dropped girder system, floor joists do not have to be cut to exact length. The joists can overlap one another and are nailed together over the girder.

To construct and install a dropped triple girder, proceed as follows:

1. *Determine the length of girder needed.*

2. *Build up the girder by nailing three layers of girder lumber together.*

   The size of lumber used will be determined by the load to be supported.

   a. *Begin assembly with one full-length board and one half-length board to stagger the end joints (Figure 35).*

      Measure joists to join over the piers. Use four 16d nails across at 12″ intervals.

   b. *Keep top edges of lumber flush.*

      Do not use lumber with large knots or a crown (slight structural bow) over ½″. Check by pulling a chalk line from end to end along the crowned edge. Any boards with a slight crown should be assembled so the crown edge will be up. The weight of the building will level the girder.

   c. *Using full-length lumber pieces, continue to build a second layer the full lengh of the girder. End with another half-length.*

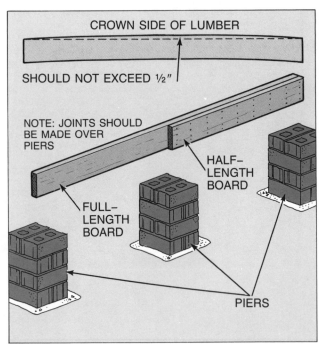

FIGURE 35. Begin a triple girder by nailing one full-length and one half-length board together.

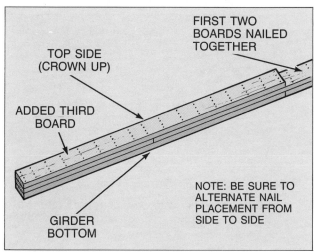

FIGURE 36. Nail the third board to the first two on the other side, making sure the joints are staggered.

d. *Turn the two nailed-together layers over and nail a third layer to the first two using 20d nails, four every 12"* (Figure 36). *Be sure to alternate placement of the nails from the other side and continue to build the third layer.*

3. *Place termite shields on the support piers or girder pockets.*

4. *Turn the girder up on its edge (crown side up) and center it on the piers or line it up with girder pockets.*

5. *Place the girder on the shortened foundation pier at each end or in the foundation girder pocket provided.*

6. *Pull a tight nylon line from the tops of each of the ends, and shim the girder to floor level.*

2. BUILDING AND INSTALLING A BUTT-TO-GIRDER

Where crawl space clearance is minimal or a dropped girder is not desirable for other reasons (such as in a basement), another girder system may be constructed. This arrangement is known as a **butt-to-girder** system. It is built to be flush with the tops of the floor joists and perimeter sill (Figure 37).

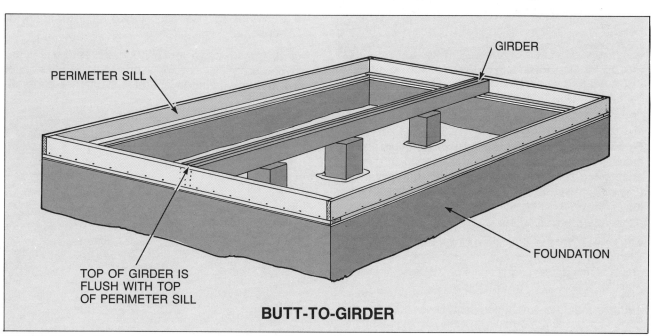

BUTT-TO-GIRDER

FIGURE 37. A butt-to-girder is constructed so that the top of the girder is even with the tops of the floor joists and the perimeter sill.

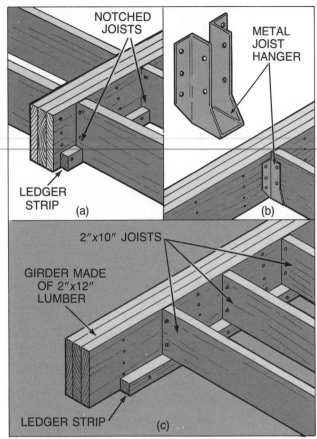

FIGURE 38. Floor joists may be joined to butt-to-girder (a) by a notched joist over a ledger strip; (b) by using a metal joist hanger; or (c) by using a wider piece of lumber and setting the joists on top of the ledger strip.

Floor joists in this system are cut to exact length and are toenailed to the girder. They may be notched and set on a plate at the bottom of the girder called a **ledger board** (Figure 38a). When metal joist hangers are used, notching of the joists is not necessary (Figure 38b).

An alternate method avoids notching floor joists by using wider pieces for the girder than for the joists (Figure 38c). When this method is used, the height of the piers and the depth of any girder pocket must be adjusted in order to maintain a level with the end joists.

To build a butt-to-girder, complete Steps 1 and 2 described for building a dropped girder. To install a butt-to-girder, proceed as follows:

3. *Place termite shields on the support piers.*

4. *Turn the girder up on its edge (crown side up) and secure at each end by nailing six 16d nails through the end joists into the end of the girder.*

   Check the position of the girder to make sure it is centered on the pier (center of the building). For a true reference, pull a nylon line taut from one end joist to the other along one edge of the girder.

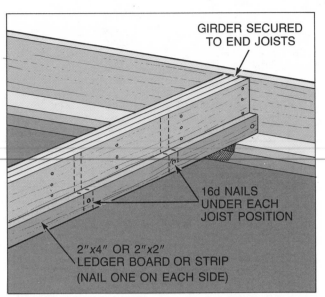

FIGURE 39. Nail the ledger strip along the bottom edge of the girder.

5. *Nail a ledger board (strip) along the bottom edges of both sides of the girder* (Figure 39).

   Use a 2" x 2" or 2" x 4" piece of the ledger strip. Use 16d nails under each floor joist position to secure the ledger strip to the girder.

A girder may be made up of only two pieces of lumber and plywood for lighter loads. Plywood should be sandwiched between the two boards for added strength (Figure 40).

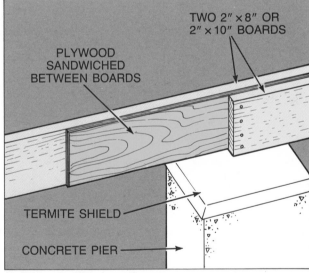

FIGURE 40. A two-piece girder with plywood sandwiched between the boards may be used for lighter loads.

STUDY QUESTIONS

1. Name the two classifications of girders.

2. What is the difference between the dead loads and live loads imposed on a floor frame.

3. Girders that support wall loads framed above the floor are called _____.

4. Floor girders are supported by _____, _____ or _____.

5. Name three types of girders that can be constructed or purchased.

6. Give two basic methods of building and installing girders.

7. One specification that often limits the use of a dropped girder is a crawlspace minimum of _____ to _____ inches.

8. Name one advantage of a dropped girder.

9. Built-up girder joints should join over _____.

10. A slight structural bow in girder stock is called the _____.

11. All floor joists must be cut to length when using a _____ girder.

12. A girder is usually placed at the _____ of the building.

13. Nail sizes used to assemble girders are _____ and _____.

14. A _____ is often placed along the bottom edge of the butt-to-girder.

15. A built-up girder of three layers of girder lumber will be _____ inches thick (nominal width).

NOTES

Floor joists are the horizontal planks, placed on edge, to which the subfloor will be nailed. In the case of two-story houses, they support the ceiling materials for the first floor and provide subflooring support for the second floor. The joists must carry all dead and live loads imposed upon them. They must be structurally strong enough to support proposed loads and rigid enough to avoid bending and vibration.

Sizes of lumber used for floor joists (2″ x 10″, 2″ x 12″, etc.), as well as the spacing between the joists and the direction they should travel, may be specified on the foundation plan. One edge of each floor joist rests upon the sill plate, while the other either overlaps or butts into a girder or overlaps an interior wall.

When selecting material for joists, consideration should be given to the allowable joist spans for various species of framing lumber. For information on some popular species, see Table IV.

Installing floor joists is discussed under the following headings:

1. Laying Out the Floor Joists

2. Cutting and Installing the Floor Joists

1. LAYING OUT THE FLOOR JOISTS

The layout of the header and girder for location of the floor joists is a most important task. The correct location of every joist is necessary if factory materials, such as plywood and particleboard, are to fit the floor joist frame. In the following example, 16″ O.C. (on-center) is used. The floor joists should be laid so that the edges of standard-sized 4′ x 8′ sheets will break over the centers of the joists. Careful planning will eliminate the unnecessary labor and waste of trimming panels.

Laying out the floor joists is discussed under the following headings:

a. Layout on Headers

b. Layout on Sill Plates

a. Layout on Headers

Proceed as follows:

1. *Begin the layout of each side from the same end of the building.*

2. *Measure 15¼″ from the outside of the end joist, and draw a line across the top edge of the header at this point* (Figure 41).

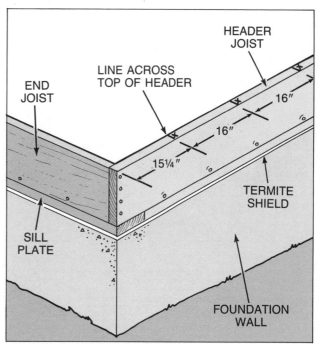

FIGURE 41. Measure 15¼″ from the outside of the end joist and mark the header.

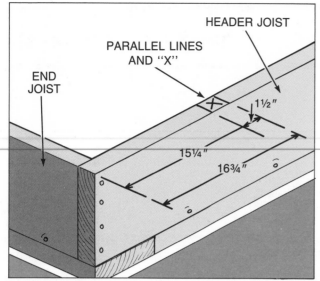

FIGURE 42. Draw a line 16¾" from the outside of the end joist, and mark an X in the area between the two marks.

3. *Place another parallel line across the edge 1½" beyond the first line (16¾" from the outside of the end joist), and place an X between these marks* (Figure 42).

The X indicates that the joist must be nailed between these marks. Two-inch rough lumber measures 1½" when surfaced. **The 1½" measure above is the actual thickness of all 2" lumber commonly used in house framing.**

4. *Drive a small nail at the 15¼" mark, or cut a small kerf (channel) in the top of the header joist along the parallel line* Figure 43).

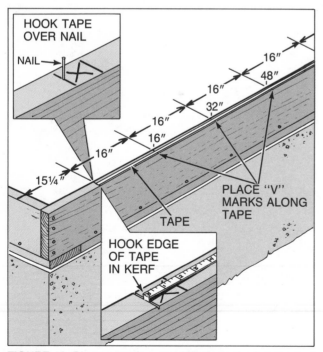

FIGURE 43. Place a mark every 16" along the header and an "x" ahead of each one.

40

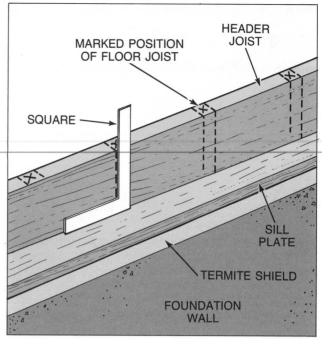

FIGURE 44. Use a square to mark the position of the floor joists on the inside of the header.

5. *Place a rule or steel tape measure, with special marks at 16" intervals, along the face of the total layout.*

Hook the tape over the nail or insert the tape's tab into the kerf, and extend the tape to the other end of the building.

6. *Place a V mark every 16" along the header joist as indicated by the tape (16", 32", 48", 64", etc.). Place an X ahead of each mark, not behind.* (Figure 43).

7. *Repeat the above steps for the other side of the house.*

On the other side, place the X behind the V mark if the joist area laps over a dropped girder.

8. *When both sides are marked, use a chalk line, pulled from side to side, to mark the center girder.*

You may use the steps given above for marking the sides if you prefer. Check your layout when you get to the other end. Correct layout should result in the same measure from the final mark to the end.

9. *Using a try square, combination square, or framing square, at 90°, draw a full line from top to bottom inside the header joist to mark the position of each floor joist* (Figure 44).

10. *Draw the joist lines for the girder.*

These steps may not be necessary for an accomplished builder, but it will help the beginning carpenter to place the joists true from top to bottom.

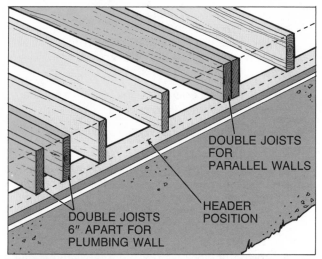

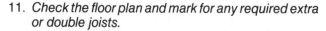

FIGURE 45. Extra joists will be required for plumbing walls and parallel walls.

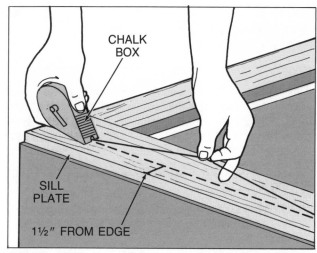

FIGURE 46. Chalk a line 1½" from the outside edge of the sill plate.

11. *Check the floor plan and mark for any required extra or double joists.*

Plumbing walls (those walls containing plumbing pipes) and parallel walls (those walls or portions that will rest on the floor running in the same direction as the floor joists) will require double joists (Figure 45).

12. *Place other double or extra joists, as needed, in areas of concentrated loads, such as kitchens and utility areas.*

13. *Lay out framing for any floor openings.*

See instructions in the "Framing Floor Openings" section for framing openings.

b. Layout on Sill Plates

Some carpenters prefer the following procedure:

1. *Before the header joists are installed, chalk a line on the sill plates 1½" from the outside edge (Figure 46).*

2. *Mark the sill plate for the joist location.*

3. *Install the joists to this line by toenailing to the sill (Figure 47).*

4. *Secure the header joist to the ends of the floor joists and the sill plate (Figure 47).*

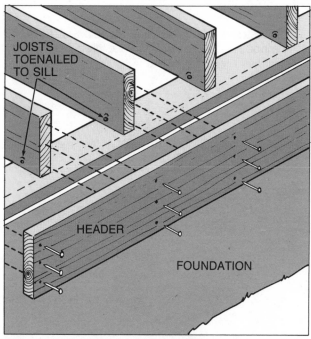

FIGURE 47. Toenail the joists in position and then nail the header to the joists and the sill plate.

2. CUTTING AND INSTALLING FLOOR JOISTS

If all layout work has been done correctly, the actual cutting and installing of the floor joists may be the easiest and quickest part of the entire job. Assuming that all work to this point is true and accurate, all floor joists can be cut by using one pattern joist. Theoretically, the joist length is the measure of one half the house width, less the thickness of the header joist and half the thickness of the girder. In any event, the actual length needed should be checked before any lumber is cut. Be especially careful to make sure of the length of the joists needed on either side of the girder.

Proceed as follows:

1. *Select a straight piece of lumber for a pattern. Square cut the end if necessary, and cut the lumber to the required length.*

   This step is not necessary if lapping joists over a dropped girder, because the end will not butt against another member (Figure 48).

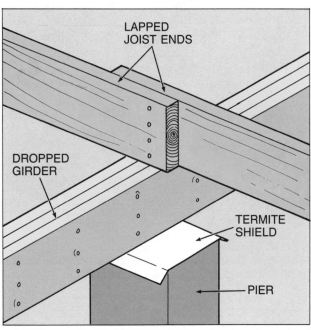

FIGURE 48. A lapped joist does not require that the ends of the joists be cut to an exact length.

2. *Place several joist pieces side by side on saw horses. Align the ends of the pattern joist and the lumber to be cut. Mark the lumber with a pencil or other marking device on the end where there is excess. Cut joists one at a time.*

   Repeat this procedure until the required number of joists have been marked and cut.

   The following is another widely used method:

   Square one end of each joist piece with a framing square, mark, and cut. With a tape rule, measure and mark the length of each joist, square with a framing square, and cut joists one at a time.

3. *Saw the lumber needed for the floor system according to the number of joist marks.*

   Any piece of stock with a crown (a slight structural bow) of more than ½" should not be used for a joist. Bad pieces of lumber can be used in shorter lengths, as for bridging, short headers, etc.

4. *Place every fifth or sixth pair of joists first in order to align the girder. Then fill in the rest as required by the layout marking (Figure 49).*

5. *Place the joists between the girder and header, crown up.*

   The building weight and deflection of the floor system will tend to straighten or level any slight crowns. A crown placed upside down may cause a low place in the floor, and a joist with too much crown may cause a hump.

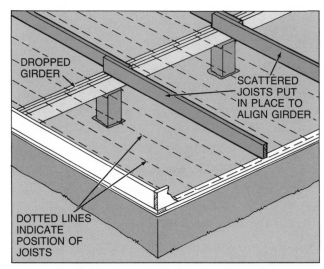

FIGURE 49. Place every fifth or sixth pair of joists first in order to align the girder.

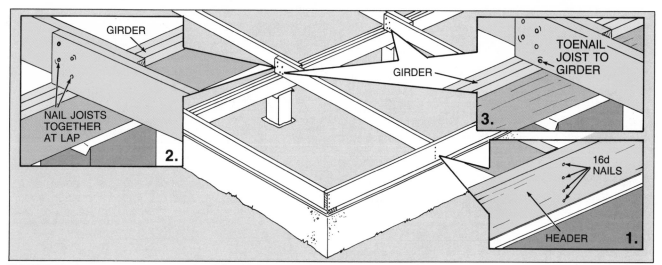

FIGURE 50. Sequence for securing joists to header, girder, and/or other joists.

6. *Nail Joists.*

   a. *If joist laps over a dropped girder, nail joist to the header first (or to sill line). Nail joist at the lap and then toenail joist to the girder (Figure 50).*

   To secure the joist, drive four 16d nails through the header into the end of the joist.

   b. *If a butt-to-girder, toenail joist to the girder first. Next, nail joist through the header. Make sure joist is flush to the top of the header and girder along the lines of the layout.*

STUDY QUESTIONS

1. The floor joists are _____ planks placed in what position _____.

2. Joists must carry all _____ and _____ loads.

3. Floor joists are usually placed _____ inches O.C.

4. An X placed beside a layout mark represents the position of a nailed _____.

5. Where are double or extra joists required in the floor system?

6. In theory, the length of floor joists will be _____ the width of the house, less the _____ of the header joist and _____ of the girder.

7. Give two methods of measuring and marking floor joists.

8. Joist stock that has excessive crown (unusuable for floor joists) can be used for _____.

9. Joists ends are nailed to the _____ and the _____.

10. Crown edges of all floor joists should be placed _____.

Bridging adds strength and rigidity to the floor frame, as well as helping to distribute concentrated loads over several floor joists. It consists of solid wood blocks or other metal or wooden pieces nailed between the joists. Correctly installed, it will help straighten and hold joists in true alignment for placement of subflooring materials.

Bridging is installed just before the subflooring materials are put down. Joist spans of over 8' should have one row of bridging, and spans of over 16' should have two rows of bridging spaced 5' or 6' apart.

Three types of bridging are discussed:

1. Solid Wood Bridging
2. Cross or Herringbone Cross Bridging
3. Metal Cross Bridging

1. SOLID WOOD BRIDGING

Solid wood bridging is the most commonly used. Some advantages are as follows:

- It offers greater rigidity and less vibrating of floor joists.

- It is cut from the same stock as joists. It is a good place to use short pieces and culled boards.

- Since the bridging blocks are installed in a staggered fashion, they can be face-nailed from both ends, resulting in a faster nailing operation.

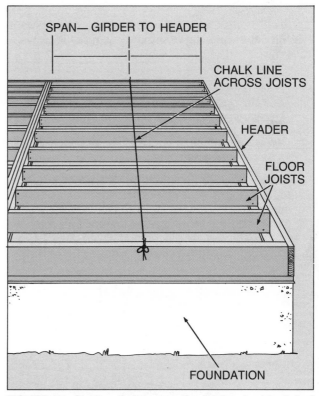

FIGURE 51. Chalk a line across the joists at midpoint of span for alignment of bridging.

Proceed as follows:

1. *Using lumber the same width as the joists, make cuts 14⅜" for 16" O.C. and 22⅜" for 24" O.C. joist spacing, assuming lumber thickness is 1½".*

2. *Chalk a line across the top of the joists at midpoint of span—header to girder (Figure 51).*

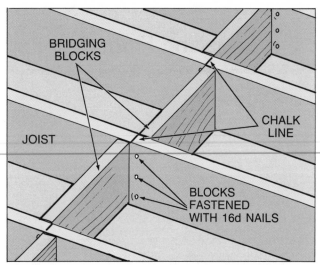

FIGURE 52. Proper alignment and nailing of solid wood bridging.

3. *Offset blocks on each side of the chalk line, and face-nail with three 16d nails at each end. Make sure blocks are flush with the top of joists* (Figure 52).

4. *Cut blocks for each end and for between double joists to required lengths.*

These measurements will be different than for the regular blocks. Always take measurements between the joists at the header for increased accuracy when measuring for these blocks (Figure 53). After the bridging blocks are placed and nailed, the joists will be joined together in straight alignment.

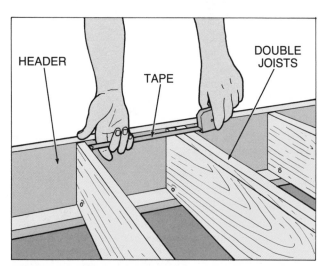

FIGURE 53. Check measurement at the header for accurate figures on the distance between joists and double joists.

NOTE: It is necessary to pull a tape from the end of the house from which the original layout was started to check the correct spacing of the joists. Oftentimes, the blocks force the joists a bit out of line if the spacing is not carefully checked and maintained. As you will see in the section on "Installing Subfloors," position of the joists at the 4' x 8' interval is critical since the edges of the 4' x 8' sheets of plywood or other panel material should meet at those points over a joist. Inaccurate positioning of these joists will cause problems with nailing and with the correct placement of additional panels. Improper joist placement could also make additional cutting of panels necessary.

2. CROSS OR HERRINGBONE CROSS BRIDGING

Cross or herringbone cross bridging consists of 1" x 4" or 2" x 4" wood installed between joists in a cross pattern as shown in Figure 54. This type of bridging is installed by toenailing cut-to-length pieces at each end with 6d or 8d nails. Normally, the tops are fastened to the joists in a staggered fashion after a chalk line has been made, as was done for solid wood bridging (Figure 51). The bottoms of the pieces are not secured until the subflooring has been put in place. If the bottoms were nailed earlier, joists could be pushed out of line.

Cross bridging has been done for years and effectively strengthens the floor structure. Labor costs, however, have reduced its popularity. The time required to cut and install the cross bridging pieces is greater than for cutting and installing solid wood bridging or metal cross bridging.

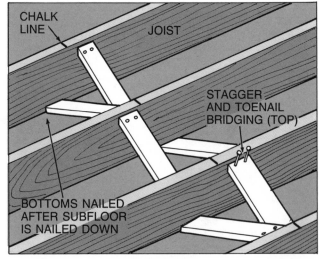

FIGURE 54. An example of herringbone cross bridging.

3. METAL CROSS BRIDGING

Another system of cross bridging uses braces made of ribbed steel instead of wood. The upper nailing flange is driven into the top of one joist, and the lower nailing flange is driven into the lower end of an adjacent joist (Figure 55).

Metal bridging may be purchased in different lengths to accommodate common joist spacing. Although the cost may be somewhat higher than for wood, the savings in labor makes them a good alternative.

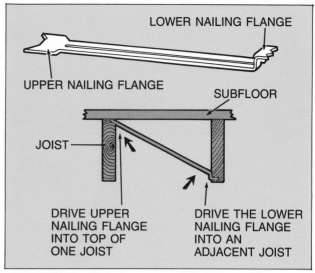

FIGURE 55. Metal cross bridging presents an opportunity to save labor costs.

STUDY QUESTIONS

1. The purpose of bridging is to add _____ and _____ to the floor frame.

2. Joists spanning 16' or over should have _____ rows of bridging spaced _____ to _____ feet apart.

3. Name one advantage of solid wood bridging.

4. Name a disadvantage of cross bridging.

5. A good alternative for wood cross bridging is the use of _____ bridging.

47

NOTES

Floor Framing over a Basement

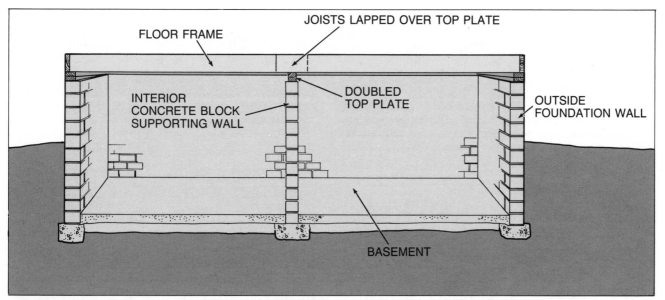

FLOOR FRAME

JOISTS LAPPED OVER TOP PLATE

INTERIOR
CONCRETE BLOCK
SUPPORTING WALL

DOUBLED
TOP PLATE

OUTSIDE
FOUNDATION WALL

BASEMENT

FIGURE 56. Floor framing over a basement is supported by foundation walls and interior walls or supports.

When building a floor frame over a basement, there are a number of factors to be addressed. Among these are consideration of adequate support for the floor frame including the girder or center support and the outside support or foundation wall (Figure 56).

Floor framing over a basement will be discussed under the following headings:

1. Providing Outside Support for Joists and Girders over a Basement

2. Providing Support for the Center Ends of Joists over a Basement

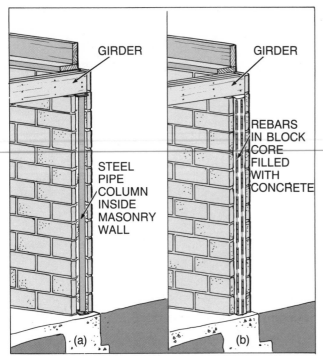

FIGURE 57. Additional support in the foundation wall may be provided by (a) a steel pipe within the masonry wall or (b) installation of rebars in the block core, filled with concrete.

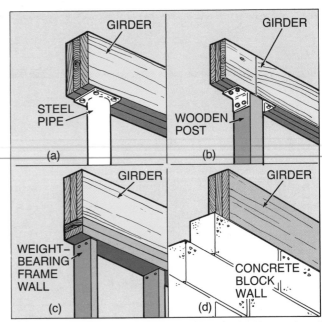

FIGURE 58. The girder for securing the center ends of joists may be supported by (a) a steel pipe; (b) a wooded post; (c) weight-bearing frame wall; or (d) a concrete block masonry wall.

1. PROVIDING OUTSIDE SUPPORT FOR JOISTS & GIRDERS OVER A BASEMENT

In general, the procedures outlined and discussed under the sections on "Installing the Perimeter Sill," "Constructing and Installing Built-Up Wood Girders," and "Installing Floor Joists," should be followed when providing outside support for joists and girders over a basement.

There are other considerations, however. Because of the additional weight of a wooden beam, wood-steel-sandwich beam, or metal beam, more support may be required in the foundation wall.

A steel pipe, known as a **lally column,** placed inside the masonry foundation built over a reinforced footing, should give adequate support (Figure 57a). The base of this pipe must be accurately placed and bolted to the footing and floor slab.

Another method of strengthening the support for a girder is the placing of **rebars** (steel reinforcing bars) in the block core and filling with concrete (Figure 57b).

Check local building codes for requirements on footings and supports.

2. PROVIDING SUPPORT FOR THE CENTER ENDS OF JOISTS OVER A BASEMENT

The center ends of the joists over a basement area may be attached to a girder supported by either steel pipes, or wooden posts (Figure 58). A weight-bearing frame wall or a concrete block wall may also support the girder to which the center ends of the joists are attached (Figure 58).

If you use a steel pipe or a wooden post, you will need a girder. This girder may be constructed of 2″ x 10″ or 2″ x

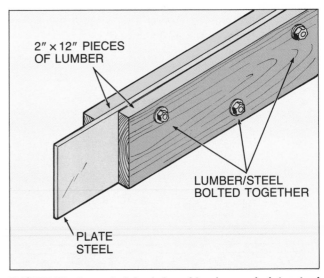

FIGURE 59. A sandwich girder of lumber and plate steel may be needed for long spans.

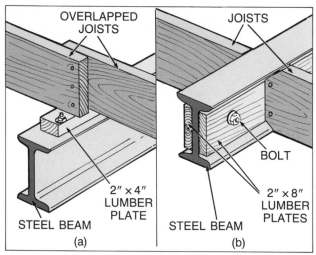

FIGURE 60. When a steel beam is used as a girder, joists may (a) overlap the beam or (b) butt against the sides of the beam

12″ wood pieces nailed together. For long spans, plate steel may be sandwiched between the wooden members (Figure 59).

A steel beam, known as a **wide flange beam** or "I" **beam,** may also be used as a support girder. Wood joists may overlap the top of the steel beam or butt against the sides of the beam (Figure 60).

Regardless of the material used for the girder, joists will need to be bridged as discussed in the other sections of this publication.

Whichever type of support you decide to use to support the girder, the floor in that area must be strengthened. The area where an interior load-bearing concrete block or frame wall is planned should have adequate additional

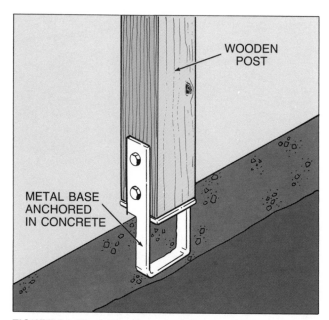

FIGURE 61. The base of any wooden post should be properly secured to the floor.

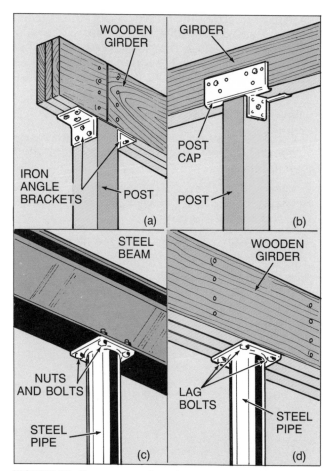

FIGURE 62. Some examples of ways pipes and posts are fastened to girders.

support beneath the finished floor. Likewise, if a steel pipe (lally column) or a wooden post is used, additional strengthening will be needed in the area where the columns will rest.

A pipe column should be attached at the base by bolts set in the top of a pier or the concrete slab. Care must be taken to assure that the bolts are accurately set into the concrete at the time the concrete is poured, or holes may be drilled into the concrete and anchors set for lay bolt fasteners.

If a wooden post is used, the width of the post should be equal to the width of the girder it supports. The base should be secured to the floor with a metal post base that is fastened to the floor or preset in concrete (Figure 61). Examples of ways in which pipe and wood columns are attached to girders, both wood and steel, are shown in Figure 62.

When a column is to be used for the support of a center girder, normally a temporary wood 4″ x 4″ or 4″ x 6″ timber is used to hold up the girder for final positioning and leveling. It is critical that the foundation wall and the top of the floor framing be level when the permanent supports are installed.

Procedures given here for building a floor framing over a basement assume that the floor frame will be constructed on the foundation wall. Additional support has been added to the wall in the area where the center girder will rest, as well as to the floor area or pier where the pipe that supports the girder will stand.

Proceed as follows:

1. *Carefully check and measure the vertical distance from the footing pad or concrete floor to the bottom of the girder.*

   Make sure that the girder is level with the top of the foundation wall for a **dropped girder** or the header joists for a **butt-to-girder.**

   Use temporary timbers to level the girder prior to installing permanent supoprts.

2. *Install pipe column(s) or wooden post(s) that have been cut to the same length as the distance measured in step 1.*

   When calculating the length, take into account any mounting attachment at the top and/or bottom of the column or post.

3. *Remove temporary supports and complete the floor framing.*

STUDY QUESTIONS

1. Give two methods of supporting a girder over a basement.

2. The outside end of each floor joist over a basement is supported by the _____.

3. One method of strengthening the outside wall support for a girder is the placing of _____ in the block core and filling with _____.

4. The foundation wall and the top of the floor framing over a basement must be _____.

5. In the case of long spans, a sandwich girder may be used, which is constructed of _____ and two _____ members.

Special framing is required for openings for fireplaces and chimneys, stairs, and in some cases, plumbing and heating ducts. The basic procedures for framing openings are about the same in most cases, all cuts must be made squarely and close measurements must be maintained to insure a good fit.

Framing for floor openings is discussed under the following headings:

1. Laying Out the Floor Opening

2. Installing Framing for Floor Opening

1. LAYING OUT THE FLOOR OPENING

To lay out the floor opening, proceed as follows:

1. *Determine the location of the rough opening.*

2. *Measure and mark the position of the inside trimmers or trimmer joists on the outside header or sill plate and the girder* (Figure 63).

   Trimmers or **trimmer joists** are joists placed on either side of the opening.

   Be sure to allow for lumber thickness when figuring the inside dimensions of the opening. If the opening is for a stairwell, the blueprint will usually show detail needed. Two points should be remembered about stairwell frame openings:

   • The length of the stairwell should be long enough to allow a minimum headroom of 6'4".

   • Stairs should be 36" or more in width.

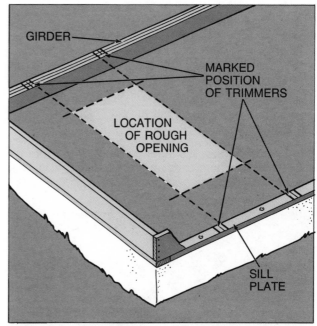

FIGURE 63. The position of the trimmers should be marked on the sill plate and the girder.

Some unusual floor framing tasks will come up in the carpentry trade. Some of these may include the following: wall projections over the foundation wall (cantilevered sections), bay windows, sunken floor areas, raised floor areas, sunken bathtubs, shower stalls, and security pockets under floor surface.

Many building plans will give details of how each job should be done. Experience will help you solve the many framing problems you will encounter.

2. INSTALLING FRAMING FOR FLOOR OPENING

To frame a floor opening that you have laid out, proceed as follows:

1. *Measure, cut and nail an inside trimmer joist on each side of the opening (Figure 64). Recheck the width of the opening.*

2. *Mark the position of the headers* (Figure 65).

   Headers should be doubled (double headers) if the opening is 4' or greater.

3. *Cut and nail in the outside header piece at each end of the opening between the trimmers using 16d nails* (Figure 66).

   The header pieces must be the same length as the width of the opening. Using a framing square, check each corner to be sure the opening is square.

4. *Mark the position of the tail joists on the headers* (Figure 66).

   Tail joists are shortened joists that run from the header to either a girder and/or supporting wall. They should follow the regular joist layout.

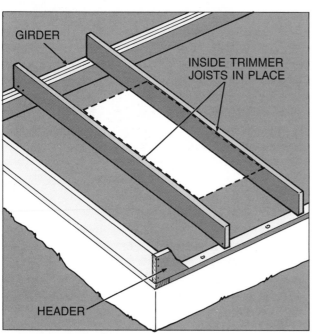

FIGURE 64. Nail the inside trimmers in place on either side of the planned opening.

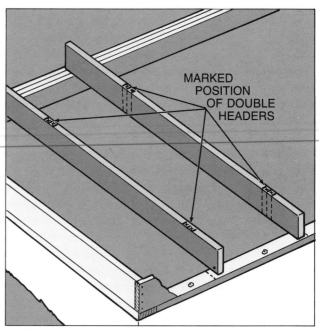

FIGURE 65. Mark the position of the headers on the trimmer joists.

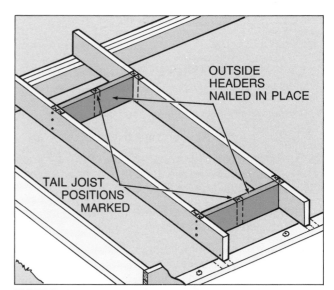

FIGURE 66. Nail the outside headers in place between the inside trimmers, and mark the position of the tail joists on the headers.

5. *Nail the tail joists to the outside headers with 16d nails. Nail through the headers into the ends of the tail joists (Figure 67).*

6. *Measure, cut, and position an inside header in place at each end of the opening, and nail each to the outside header creating a double header (Figure 68).*

7. *Using 16d nails, double the trimmer joists by nailing an additional joist on the outside of both trimmer joists (Figure 69).*

If a regular floor joist falls close to the trimmer location, install the regular joist after the frame for the opening is completed. This will make construction of the opening easier by providing additional space for nailing.

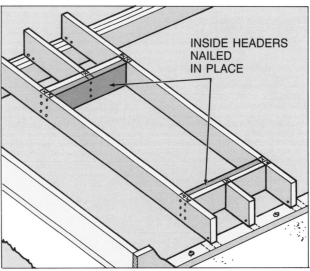

FIGURE 68. Nail the inside header pieces to the outside header pieces.

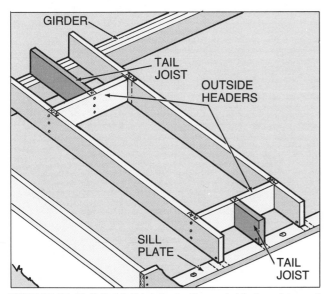

FIGURE 67. Nail the tail joists to the outside header, and toenail them to the sill plate and the girder.

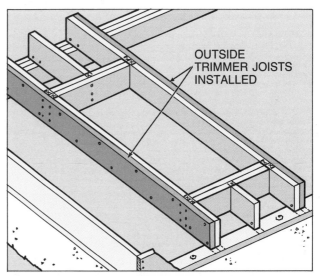

FIGURE 69. Double the trimmer joists by nailing joists to the outside of each trimmer joist.

STUDY QUESTIONS

1. Special openings have to be framed in floors to accommodate: _____.

2. When framing openings, how should cuts and measurements be made to insure a good fit?

3. _____ are joists placed on either side of a floor opening.

4. The minimum stairwell opening is _____ inches wide.

5. Minimum headroom for a stairwell is _____.

6. Give three special floor framing tasks that may have to be done by the carpenter.

7. What are tail joists?

8. List the following components in order of assembly in a floor opening—outside header, inside trimmer, tail joist, outside trimmer, inside header.

The subfloor is the lumber, plywood, or nonveneered panels, such as waferboard, nailed to the floor frame. The subfloor strengthens the entire floor frame and forms a base for the finish floor material. It is also important because the walls of the structure are laid out, framed, and raised into place on top of the subfloor.

NOTE: The selection and installation of floor underlayment—the paneled material that is placed on the subfloor to provide a smooth, even surface for finished floor materials—is not discussed in this publication.

At one time, board lumber was the most widely used material for subflooring; however, the majority of today's builders use plywood. There is a tremendous savings in labor costs when panels are used instead of boards. The most widely used panel product for subflooring in residential construction is plywood. In recent years, other nonveneered panels have grown in popularity and their use is increasing. A discussion on plywood and non veneered panel products can be found in the section on "Plywood and Other Wood Panel Products."

Regardless of the panel material selected for subflooring, the installer must make sure that the proper grade and thickness are used to meet specifications and satisfy local building codes.

Installing subflooring is discussed under the following headings:

1. Installing Plywood Subflooring

2. Installing Other Panel Products

1. INSTALLING PLYWOOD SUBFLOORING

Plywood panels can be placed quickly and easily, with a minimum of sawing. Before you begin, there are several things to consider.

First, subflooring panels are placed over the floor joists so that the long sides of the panels run at a right angle (perpendicular) to the joists. Second, joints are always staggered (Figure 70).

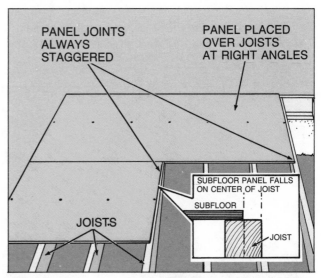

FIGURE 70. Panels are always placed at right angles to the joists and joints are always staggered.

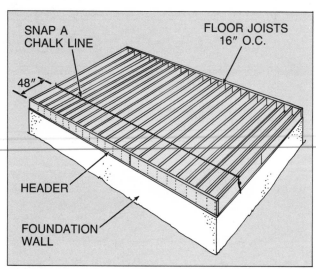

FIGURE 71. Chalk a line across the joists 4' from the edge of the header.

Proceed as follows:

1. *Beginning at either side of the floor frame, measure in 4' (48") from each corner and chalk a line on top of the joists the full length of the building (Figure 71).*

2. *Starting at the end where the joist layout began, lay a full 4' x 8' panel along the chalk line, square with the outside header (Figure 72).*

   Secure the position of the panel by nailing an 8d nail along and 1" in from the edge of the panel into the center of each joist (Figure 72).

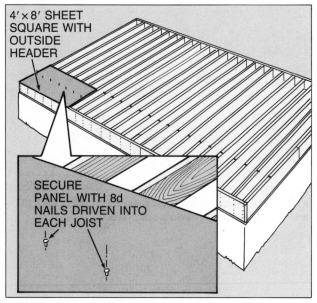

FIGURE 72. Lay the first panel along the chalk line square with the header and end joists.

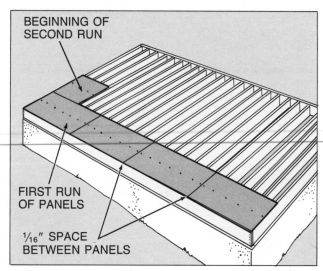

FIGURE 73. Lay panels end to end along the header and the chalk line, leaving 1/16" space between the ends of the panels.

3. *For the first run, lay panels end to end until reaching the end of the floor frame. Space the ends of the panels 1/16" apart (Figure 73).*

   The 1/16" spacing is provided to allow for expansion of the panels, which may be wet before dry-in is completed.

4. *Start the second run with a half panel (4' x 4') to create the stagger pattern (Figure 73). Leave ⅛" space between sides of the panels in the first and second runs.*

   A 16d nail can be used as a guide for this spacing. The ends of the panels must be staggered a minimum of 32" (2 joists, 16" O.C.).

   The face grain of half sheets of plywood (4'x4') should be laid so that the grain runs in the same direction as adjacent panels.

   NOTE: If you wish, you may lay pieces of 96", 64" and 32" side by side to start the first three runs. Use of these starter pieces will insure that the joints will be staggered throughout the subflooring process.

5. *Continue the second run by laying full 4' x 8' panels parallel with the first run.*

6. *As the subflooring panels are placed, cut out any openings framed in the floor.*

7. *Continue the procedure described in steps 2 through 5 for each additional run until all the panels are in place (Figure 74).*

58

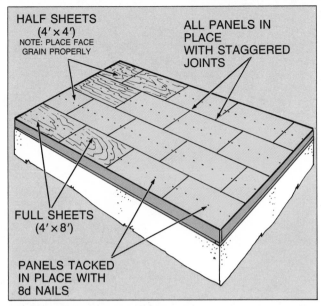

FIGURE 74. Continue to put down all panels until the floor frame is completely covered.

8. *After all panels are in place and tacked, locate the center of the first joist and snap a chalk line across the tops of the panels (Figure 75). This line will be your guide for nailing into the center of the joists.*

Since wall locations are indicated on the subflooring by chalk lines, some carpenters prefer to use an alternate method to provide a nailing guide using a line strung between two mason blocks. The line should be pulled taut over the center of a given joist, from header to header. Nail along the line using the line as a guide. Repositioning of this nailing guide is easy to accomplish by sliding the blocks along the headers.

9. *Using the chalk line (or other line) as a guide, nail an 8d nail every 6"-8" along the panel edges (where the edges fall over a joist) and at 12" O.C. along the floor joists. If a glued floor system is used, a nail spacing of 12" O.C. may be used throughout the installation process. Be sure all nails go into the frame members.*

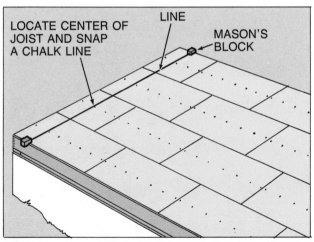

FIGURE 75. Use a chalk line or mason's block and line as a nailing guide.

FIGURE 76. Pneumatic guns speed the fastening of sub-flooring material to the flooring frame.

Actual nailing of the panels can be done by the less skilled members of the work crew, using hammers or a pneumatic gun (Figure 76).

Nails at end joints may be slightly angled into the joists, and all nails must be driven down tight.

10. *Chalk any ends or edges not previously cut, and trim off excess with a power hand saw.*

2. INSTALLING OTHER PANEL PRODUCTS

In recent years, panel products made of wood particles, flakes, or strands bonded together and formed into sheets have gained in popularity as material for subflooring. Since an adhesive or **glued floor system** is probably the most common method of applying these structural panel floors, there are special considerations that must be taken into account.

The procedures already outlined for installing plywood should be followed with one basic difference. In this system, each panel must be nailed down *completely* before moving on to installing the next panel. Since the adhesive is placed onto the tops of the joists just before laying the panel down, panels must be nailed down immediately to prevent the adhesive or glue from "skimming over" and preventing proper bonding (Figure 77). This "skim-over period" will vary depending on the adhesive being used and the temperature at the time of installation. In many cases, skimming over will occur in 10 or 15 minutes.

59

This combination of adhesive and nailing has proven effective in the elimination of squeaks and further stiffens the entire floor system.

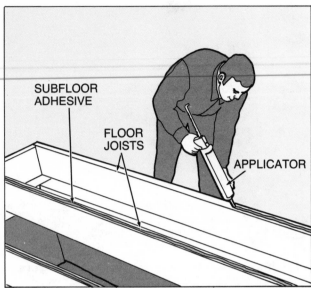

SUBFLOOR
ADHESIVE

FLOOR
JOISTS

APPLICATOR

FIGURE 77. Subfloor adhesive is applied to the tops of floor joists before the subflooring panels are nailed in place.

STUDY QUESTIONS

1. The subfloor forms a _____ for _____ floor materials.

2. The most widely used material for subflooring today is_____.

3. End joints on subfloor panels should be staggered a minimum of _____ inches.

4. Plywood is nailed to the floor frame using _____ (size) nails.

5. Panels on subfloor should be spaced _____ apart on ends and _____ apart on sides.

6. Nails should be placed across panels in each joist every _____ to _____ inches.

7. How can a beginner carpenter assure that the nails go into the joists?

8. What advantages does a construction adhesive provide when used in subfloor construction?

9. Why do some carpenters not want to snap chalk lines as a guide for nailing?

10. Why is the spacing between panels important?

11. Why is it important to nail down a glued panel completely before moving on to another panel?

ABBREVIATIONS

The following abbreviations are commonly used in today's lumber industry. There may be variations, such as in the use of all capital letters or the use or deletion of periods. This list is not complete but does provide a working knowledge of abbreviations you may deal with on a day-to-day basis.

AD — air-dried
ALS — American Lumber Standards
AWPB — American Wood Preservers Assocation
avg — average
B & B — B and better
B & S —beams and stingers
bd. — board
bd. ft. — board foot
b.m. — board measure
Bev. — beveled
Clr. — clear
Com — common
Dim — dimension
Dkg — decking
Fac — factory
FOK — free of knots
Frt. — freight
ft. — foot or feet
G — girth
in — inch
Ind — industrial
J & P — joints and planks
KD — kiln-dried

Lft. — linear foot or feet
Lin. — linear
M — thousand
MBM — thousand (feet) board measure
m.c. — moisture content
Mft. — thousand feet
No. — number
O.C. — on center
Part. — partition
pcf — per cubic foot
Pcs. — pieces
P & T — posts and timbers
Reg. — regular
Rgh — rough
R/L — random lengths
SD — grade stamp indicating lumber seasoned to a moisture content of 19 percent or less.
SE — square edge
S-GRN — grade stamp indicating lumber shipped unseasoned or "green"
Sel. — select
Specs. — specifications
Std. — standard
Std. lgths — standard lengths
STR — structural
SYP — Southern Yellow Pine
S1S — surfaced on one side
S2S — surfaced on two sides
S4S — surfaced on four sides
wdr. — wider
Wt. — weight
YP — Yellow Pine

NOTES

A

Anchor bolt 27
Architect's design 19

B

Basement, framing over 49
    support for center ends 50
    support for girders 49
    support for joists 49
Bridging 17, 21, 42, 45
    cross 42
    herringbone 46
    metal cross 47
    solid wood 45, 46
Building permit 5

C

Chemically treated lumber 13
Composite board 12
Cross laminating 11

E

Estimating lumber 21
    floor framing 17
    plywood 15
    subfloors 22

F

Floor openings
    framing 53
    installing framing 54
    laying out 53

G

Girders 17, 21
    built up 33
    butt-to-girders 35, 36, 51
    dropped girder 33, 34, 51
    pocket 34, 35
Glued floor system 59

H

Header 21, 54

I

I beam 50

J

Joists 17, 19, 21, 31, 34, 39, 55
    cutting 42
    installing 42
    layout headers 39
    layout sill plates 41
    tail 54, 55
    trimmer 53, 54

L

Lally column 49, 51
Ledger strips 19, 20, 35, 36
Lumber
    air dried 7
    boards 9, 57
    chemically treated 13
    cost 10, 19
    dimension 9
    figuring 9, 10
    finish materials 9
    graded 7, 8
    kiln dried 7
    measuring 9, 10
    plywood 10
    wood panel products . . . 11, 12, 13

M

Materials availability 19
Materials selection 19
    bridging 20
    ledger strips 20
    sills and joints 20
    subfloors 22

N

Nails and fasteners 14, 15

O

Oriented strand board 13

P

Panel grading 11
Particleboard 12, 23
Perimeter sill 31
Pipe column 51, 52
Platform framing 17
Plywood 10, 11, 22, 57
Pressure treated 13

R

Rebar 50

S

Screws 15
Sill plates 25, 26, 28, 31, 32, 41
Softwood
    board 9
    dimension 9, 10
    grading systems 7
    timbes 9
Specifications 19
Staples 15
Subfloors 22, 17, 39, 57
    other panel products . . . 12, 59, 60
    plywood 57, 58

T

Termite shields 25, 28, 34, 36
Toenailing 32, 46
Trimmers 53, 54
Two-story floor framing 18

V

Veneers 11

W

Waferboard 12, 57

Acknowledgements

PICTURE CREDITS

Appreciation is hereby expressed to the following companies and individuals who graciously supplied photographs for use in this publication. Company names are followed by a listing of the particular figure number(s) supplied.

Companies and Institutions

American Plywood Assocation, 7011 South 19th Street, Tacoma, Washington 98411
Figure 10

Duo-Fast Corporation, 3702 River Road, Franklin Park, Illinois 60131
Figures 15, 76

Koppers Company, Inc., 1050 Koppers Building, Pittsburgh, Pennsylvania 15219
Figures 11, 12a

National Pest Control Association, 8100 Oak Street, Dunn Loring, Virginia 22027
Figure 21

CONTRIBUTORS OF LITERATURE AND OTHER MATERIALS

Companies and Institutions

American Lumber Standards Committee, P.O. Box 210, Germantown, Maryland 20874

American Wood Preserver's Association, P.O. Box 849, Stevenville, Maryland 21666

Georgia-Pacific Corporation, 133 Peachtree Street, N.E., Atlanta, Georgia 30348

National Particleboard Association, 18928 Premiere Court, Gaithersburg, Maryland 20879

Southern Building Code Congress, 900 Montclair Road, Birmingham, Alabama 35213

Southern Forest Products Association, P.O. Box 52468, New Orleans, Louisiana 70152

Southeastern Lumber Manufacturers Association, P.O. Box 1788, Forest Park, Georgia 30061

Southern Pine Inspection Bureau, 4709 Scenic Highway, Pensacola, Florida 32504

Southern Pressure Treaters Association, P.O. Box 617, Brewton, Alabama 36427

Western Wood Products Association, Yeon Building, 522 S.W. Fifth Avenue, Portland, Oregon 97204